Energetic Materials and Munitions

Energetic Materials and Munitions

Life Cycle Management, Environmental Impact and Demilitarization

*Edited by
Adam S. Cumming
Mark S. Johnson*

Editors

Prof. Adam S. Cumming
University of Edinburgh
School of Chemistry
David Brewster Road
Joseph Black Building
EH9 3FJ Edinburgh
United Kingdom

Dr. Mark S. Johnson
US Army Public Health Center
5158 Black Hawk Rd
Aberdeen Proving Ground
21010 Gunpowder MD
United States

All books published by **Wiley-VCH** are carefully produced. Nevertheless, authors, editors, and publisher do not warrant the information contained in these books, including this book, to be free of errors. Readers are advised to keep in mind that statements, data, illustrations, procedural details or other items may inadvertently be inaccurate.

Library of Congress Card No.: applied for

British Library Cataloguing-in-Publication Data
A catalogue record for this book is available from the British Library.

Bibliographic information published by the Deutsche Nationalbibliothek
The Deutsche Nationalbibliothek lists this publication in the Deutsche Nationalbibliografie; detailed bibliographic data are available on the Internet at <http://dnb.d-nb.de>.

© 2019 Wiley-VCH Verlag GmbH & Co. KGaA, Boschstr. 12, 69469 Weinheim, Germany

All rights reserved (including those of translation into other languages). No part of this book may be reproduced in any form – by photoprinting, microfilm, or any other means – nor transmitted or translated into a machine language without written permission from the publishers. Registered names, trademarks, etc. used in this book, even when not specifically marked as such, are not to be considered unprotected by law.

Print ISBN: 978-3-527-34483-3
ePDF ISBN: 978-3-527-81664-4
ePub ISBN: 978-3-527-81666-8
oBook ISBN: 978-3-527-81665-1

Cover Design: Adam-Design, Weinheim, Germany
Typesetting   SPi Global Private Limited, Chennai, India
Printing and Binding   CPI books GmbH, Germany

Printed on acid-free paper

10 9 8 7 6 5 4 3 2 1

Contents

Preface *xi*

1 Introduction and Overview *1*
Adam S. Cumming
1.1   Introduction *1*
1.2   Legislative Impact *2*
1.3   NATO Studies *4*
1.4   New Ingredients and Compositions *5*
1.5   Toxicology *6*
1.6   Life-Cycle Analysis *7*
1.7   Managing Contamination and Clean-Up *7*
1.8   Disposal Now and in the Future *8*
1.9   Recycling *8*
1.10 Conclusions *9*
       References *9*

2 General Introduction to Ammunition Demilitarization *13*
David Towndrow
2.1   Part one – Logistics, Costs, and Management *13*
2.1.1   Introduction *13*
2.1.2   Context of Demilitarization *14*
2.1.2.1   The Scale of the Issue *14*
2.1.2.2   Factors Influencing Demilitarization *15*
2.1.3   Demilitarization Process *17*
2.1.3.1   Basic Stages of Demilitarization *17*
2.1.3.2   Demilitarization Facilities *20*
2.1.4   Demilitarization Techniques *20*
2.1.4.1   Demilitarization Techniques and Processes *20*
2.1.4.2   Maturity and Use of Demilitarization Techniques *21*
2.1.5   Summary *26*
2.2   Part Two – Environmental Issues and Demilitarization *27*
2.2.1   Introduction *27*
2.2.2   Demilitarization Process *27*
2.2.2.1   Technical and Environmental Issues *27*
2.2.2.2   Open Burning (OB) and Open Detonation (OD) *29*

2.2.2.3	Open Burning	*32*
2.2.2.4	Open Detonation	*33*
2.2.2.5	Examples of Cost and CO_2 in Demilitarization Options	*36*
2.2.3	Design for Demilitarization (DFD)	*40*
2.2.3.1	NATO AOP 4518 (Revised 2018)	*40*
2.2.3.2	A Munition Manager's Perspective of Disposal Plans	*44*
2.2.3.3	Future Trends	*45*
	References	*46*

3 Assessment and Sustainment of the Environmental Health of Military Live-fire Training Ranges *47*
Sonia Thiboutot and Sylvie Brochu

3.1	Introduction	*47*
3.2	Background and Context	*48*
3.3	Munition-Related Contaminants	*51*
3.4	Surface Soil Characterization in Live-fire Training Ranges	*52*
3.4.1	Safety Aspects	*53*
3.4.2	Data Quality and Sampling Objectives	*53*
3.4.3	Importance of Soil Sample Processing to Ensure Representativeness	*56*
3.4.4	How Clean is Clean?	*57*
3.4.5	Risk to the Receptors Through the Transport of Munitions Constituents	*58*
3.5	Methodology for the Precise Measurements of MC Sources	*61*
3.5.1	Explosive Footprints in Impact Areas	*61*
3.5.2	Firing Positions	*64*
3.6	Tailored Management Practices: Mitigation and Remediation	*67*
3.6.1	Mitigation Measures	*67*
3.6.1.1	Analytical Tool and Adsorption Method for MCs in Aqueous Samples	*67*
3.6.1.2	Thermal Treatment of Shoulder Rocket Propellant-Contaminated Surface and Subsurface Soils	*68*
3.7	Emerging Constituents	*69*
3.8	Conclusion	*70*
	References	*71*

4 Greener Munitions *75*
Sylvie Brochu and Sonia Thiboutot

4.1	Background and Context	*75*
4.2	Munitions Constituents of Concern	*77*
4.3	Source of Munitions Constituents	*78*
4.4	Greener Munitions Development Approach	*79*
4.5	RIGHTTRAC	*82*
4.5.1	Energetic Formulation Selection	*83*
4.5.1.1	Main Explosive Charge	*83*
4.5.1.2	Performance	*83*
4.5.1.3	IM Properties	*83*
4.5.1.4	Fate, Transport, and Toxicity	*84*

4.5.2	Main Propellant Charge	*86*
4.5.2.1	Performance	*86*
4.5.2.2	Modular Charges	*87*
4.5.2.3	IM Properties	*87*
4.5.2.4	Fate, Transport, and Toxicity	*88*
4.5.3	Field Demonstration	*89*
4.5.3.1	Final Selection	*89*
4.5.3.2	Gun Testing	*89*
4.5.3.3	Detonation Residues	*90*
4.5.4	Life-Cycle Analysis	*91*
4.5.5	Summary	*92*
4.6	New Enhanced and Green Plastic Explosive for Demolition and Ordnance Disposal	*92*
4.6.1	PETN Option	*93*
4.6.1.1	Performance	*93*
4.6.1.2	Deposition Rate	*94*
4.6.1.3	Fate, Transport, and Toxicity	*94*
4.6.2	HMX Option	*95*
4.6.3	Summary	*96*
4.7	Conclusions	*96*
	References	*98*
5	**Pyrotechnics and The Environment**	*103*
	Ranko Vrcelj	
5.1	Introduction	*103*
5.2	Registration, Evaluation, Authorisation and Restriction of Chemicals (REACH)	*105*
5.3	Qualification	*107*
5.4	Civilian Studies	*107*
5.5	Production	*109*
5.6	Site Location	*110*
5.7	Production	*112*
5.8	Raw Materials Acquisition and Quality Control	*112*
5.9	Specific Materials Production	*114*
5.10	Heavy Metals	*115*
5.11	Perchlorates and Chlorates	*116*
5.12	Smokes	*116*
5.13	Volatilization Smokes	*116*
5.14	Magnesium Teflon Viton (MTV) Countermeasures	*116*
5.15	Resins, Binders, and Solvents	*117*
5.16	Storage	*117*
5.17	Packaging Waste	*118*
5.18	Usage and Disposal	*118*
5.19	Heavy Metals	*118*
5.20	Perchlorates and Chlorates	*122*
5.21	Smokes	*124*
5.21.1	Obscurant Smokes	*124*
5.21.2	Volatilization Smokes	*124*

5.21.3	MTV *125*
5.22	Disposal and Waste Burning *126*
5.23	The Future? *127*
5.24	Suitably Qualified and Experienced Person (SQEP) Issues *128*
5.25	Integration *129*
	Acknowledgements *133*
	References *133*

6	**Munitions in the Sea** *139*
	Sandro Carniel, Jacek Beldowsky, and Margo Edwards
6.1	Introduction *139*
6.2	The Controlling Factors *141*
6.2.1	Environmental Aspects *141*
6.2.2	Corrosion *142*
6.2.3	Fate and Transport of Constituents *144*
6.2.4	Sea-Disposal Process *145*
6.3	Tools for Assessment and Remediation *145*
6.3.1	Acoustic Sensors *145*
6.3.2	EM Sensors *146*
6.3.3	Optical Sensors *146*
6.3.4	Platforms *146*
6.3.5	Navigation and Positioning *147*
6.3.6	Remediation *149*
6.4	The Outstanding Problems *150*
6.4.1	Technical Aspects *150*
6.4.1.1	Location *150*
6.4.1.2	Detection *150*
6.4.1.3	Monitoring *151*
6.4.1.4	Handling *151*
6.4.2	Environmental Aspects *152*
6.4.2.1	Chemical Degradation of MEC *152*
6.4.2.2	Long-Term and Long-Distance Transport *153*
6.4.2.3	Ecotoxicological Aspects *154*
6.4.3	Geopolitical Aspects *156*
6.5	Moving Forward *159*
6.5.1	Global Collaboration *159*
6.5.2	Recent Global EU and NATO Efforts *160*
6.5.3	Advantages of Joint Efforts *161*
	Glossary *162*
	Acknowledgements *163*
	References *163*

7	**Environmental Assessment of Military Systems with the Life-Cycle Assessment Methodology** *169*
	Carlos Ferreira, Fausto Freire, and José Ribeiro
7.1	Overview of the Life-Cycle Assessment Methodology *170*
7.1.1	Life-Cycle Thinking *170*

7.1.2	Life-Cycle Assessment	*171*
7.1.3	Purpose of Life-Cycle Assessment Studies	*173*
7.2	The Four Phases of the LCA Methodology Applied to a Case Study	*174*
7.2.1	Goal and Scope	*174*
7.2.1.1	Functional Unit	*175*
7.2.1.2	System Boundaries	*176*
7.2.2	Life-Cycle Inventory	*178*
7.2.3	Life-Cycle Impact Assessment	*182*
7.2.3.1	Life-Cycle Impact Assessment Methods	*185*
7.2.3.2	Life-Cycle Impact Assessment Software	*187*
7.2.3.3	Life-Cycle Impact Assessment of the Case Study	*188*
7.3	Limitations of Life-Cycle Assessment	*194*
7.4	Conclusions	*194*
	References	*195*

8	**Integrating the 'One Health' Approach in the Design of Sustainable Munition Systems**	*199*
	Mark S. Johnson	
8.1	General Background	*199*
8.2	Munition Compounds and Aetiology of Environmental, Safety, and Occupational Health Issues: Lessons Learnt	*199*
8.3	Core Operational ESOH Data: Needs and Requirements	*200*
8.3.1	Life Cycle Environmental Assessment	*200*
8.3.2	Bridging Communication Between Research and Acquisition	*200*
8.3.3	ESOH Data Requirements	*201*
8.3.3.1	Approaches	*201*
8.4	Current and Evolving Regulatory Interests	*207*
8.5	Case Studies and Cost Analysis	*207*
8.5.1	M116, 117, 118 Simulators	*207*
8.5.2	M-18 Violet Smoke	*208*
8.5.3	Cost and Time Considerations	*208*
8.6	Summary	*210*
	Acknowledgements	*210*
	References	*210*

9	**Overview of REACH Regulation and Its Implications for the Military Sector**	*213*
	Carlos Ferreira, Fausto Freire, and José Ribeiro	*213*
9.1	Introduction	*213*
9.2	Regulation for Hazard Substances	*214*
9.2.1	Overview of Previous Legislation Concerning Hazard Substances in the European Union	*214*
9.2.2	Overview of REACH Regulation	*215*
9.2.3	Discussion of REACH Regulation	*217*
9.3	Conclusions	*225*
	References	*225*

Contents

10 Development and Integration of Environmental, Safety, and Occupational Health Information 227
Mark S. Johnson
10.1 Introduction 227
10.2 Phased Approach to a Toxicology Data Requirement 228
10.3 Research, Development, Testing, and Evaluation 228
10.3.1 Conception 228
10.3.2 Synthesis 231
10.3.3 Testing/Demonstration 232
10.3.4 Acquisition 233
10.3.5 Engineering and Manufacturing 234
10.3.6 Demilitarization 235
10.4 Other Data Requirements 235
10.4.1 Environmental 235
10.4.1.1 Fate and Transport 235
10.4.1.2 Ecotoxicity 237
10.4.1.3 Field Monitoring 237
10.4.1.4 Disposal 237
10.4.1.5 Occupational – Industrial Hygiene 237
10.4.2 Regulatory 238
10.4.2.1 Toxic Substance Control Act 238
10.4.2.2 REACH 238
10.4.3 Integrating Weight-of-Evidence into Decision-Making 238
10.5 Concluding Remarks 238
Acknowledgements 239
References 239

11 Research Priorities and the Future 241
Adam S. Cumming
11.1 Introduction 241
11.2 Greener Munitions 242
11.3 Studies and Their Effect 243
11.4 The Problems and the Changing Requirements 245
11.4.1 Land Management and History 246
11.5 Security Issues and Their Impact on Requirement 247
11.6 Future Options and Needs in a Changing Political Landscape 247
11.7 Conclusions 250
References 251

Index 253

Preface

Concern for the environment is important in all aspects of science at present. Energetic materials and munitions are by design materials that result in environmental releases. While sustainable use of these materials is essential worldwide to meet defence and security needs, they are not exempt from regulations that are aimed at protecting public health. Therefore, mitigating past contamination, and minimising present and future impact while munitions are used will be of enduring concern as long as these materials are employed.

There have been several publications, both as books or scientific papers that have addressed aspects of the problem, but no overview has been produced aimed at drawing the various strands together.

This book is therefore intended to be an introduction to the wide area of environmental aspects as they affect munitions and energetics. It will introduce scientists, technologists and users to the understanding of environmental issues for munitions and of ways to manage potential risks. It is aimed at providing a basic understanding of the science and its application to reducing environmental risks in the design, use and disposal of munitions. It therefore provides chapters covering the various topics and considerations which will be shown to affect each other and to affect planning for the future.

The research, development, testing, production, use, and disposal of these munitions contributes to the overall environmental impact. Research and testing involves environmental releases, as do products from manufacturing and demilitarisation. Since handling of munitions with energetic materials requires great care and considerable cost, the approach to demilitarisation is covered with examples of the types of problems currently found and an assessment of future needs and directions.

The environmental impact of the processes must be acceptable to an increasingly critical general population to avoid anti-military backlash or unintelligent imposition of any environmental law. Clean up and restoring areas where military activities have resulted in contaminated ground or water often requires significant resources. Past practices such as dumping at sea, open burning, or dumping into land-fill sites are no longer generally acceptable. There is also a need to understand and minimise the environmental impact from munitions, so we can handle and manage that impact properly.

The chapters cover Demilitarisation; Toxicology; Greener Munitions of all kinds; Land Management and the problem of Underwater contamination as well as Life Cycle Analysis and also discusses options for the future through coverage of current research and how this might affect the application of technology.

It arose from a series of lectures sponsored by the NATO Science and Technology Organisation aimed providing just that overview and has been expanded and updated to more fully provide a useful review text, which should lead to deeper study where appropriate. The authors are international experts with wide experience in international collaborative work in the area and have drawn on that experience for their chapters.

We are grateful to our co-authors for their time, support and contributions and to NATO STO for both providing the support for the initial lectures and agreeing to its development into this volume

Adam S. Cumming
Mark S. Johnson

1

Introduction and Overview

Adam S. Cumming

University of Edinburgh, School of Chemistry, Joseph Black Building, The King's Buildings, David Brewster Road, Edinburgh, EH9 3FJ, UK

1.1 Introduction

Armed forces countries possess and use large quantities of munitions. Civil authorities, such as space agencies, also use quantities of energetic materials. The production, use, and disposal of these materials make a contribution to the overall environmental impact. Handling of munitions with energetic materials requires great care and considerable cost. The environmental impact of the processes must be acceptable to an increasingly critical general population to avoid public concern and be acceptable under environmental laws. Significant funds must be used to clean up and restore areas where military activities have polluted the ground or water. Past practices such as dumping at sea or into landfill sites are no longer generally acceptable. There is a need to know and minimize the environmental impact from munitions so that environmental management can be undertaken properly.

Governments have a duty of care to the members of their armed forces, and all reasonable precautions must be exercised to ensure safe use of munitions. For example, some weapons systems can spread over 70% of their energetic material, particularly propellant around the shooting range. This is a health risk with the hazard of fires after prolonged use of the shooting range and there is also a work environment hazard. It is also an environmental hazard since a propellant's environmental hazard assessment is usually based on the final combustion products and not on the propellant itself.

The design of new weapons should include disposal procedures and an environmental impact statement. The understanding of munitions disposal is still lagging behind this design requirement although progress has been made, as is noted in this volume. However, to better meet the requirement, it is important to fully understand the environmental issues so that they do not place undue constraints on the design of weapons. Such understanding can also reduce the costs.

Energetic Materials and Munitions: Life Cycle Management, Environmental Impact and Demilitarization, First Edition. Edited by Adam S. Cumming and Mark S. Johnson.
© 2019 Wiley-VCH Verlag GmbH & Co. KGaA. Published 2019 by Wiley-VCH Verlag GmbH & Co. KGaA.

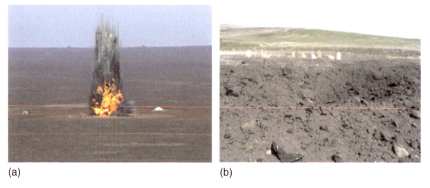

Figure 1.1 Demonstration of (a) a large detonation and (b) the aftermath – residues left.

To be able to assess the environmental impact of the munitions, we need the right environmental assessment tools. To minimize the impact of manufacture and manage green munitions, it is important to look at all processes governing these activities.

This activity has been developing for many years and has been reported [1–6].

Finally, there is the need to understand, manage, and decontaminate after events such as those mentioned subsequently (Figure 1.1).

What we need to develop is a planned management method, and this is discussed later (Figure 1.2).

1.2 Legislative Impact

Public pressure has led to the implementation of legislation to manage environmental impact. This has gradually evolved from *ad hoc* national approaches to systematic regulations such as the Registration, Evaluation, Authorisation and Restriction of Chemicals (REACH) in the European Union (EU) where the law is limiting and controls the availability and use of materials.

While such legislation is of prime importance in the nations where it is directly applied, it has an effect elsewhere since import and export of materials is transnational and those imposing the legislation are usually the largest users and hence the largest market for the materials. For example the imposition of REACH terms affects the sales of energetic materials, etc. to EU nations from outside the EU [7].

The US Environmental Protection Agency (EPA) and the EU [7] have focused on minimizing impact, and in the EU legislation the control of chemicals is being introduced. Therefore, changing public perception and new legislation means that the environmental impact of munitions and their ingredients cannot be ignored. We require understanding of the problems if they are to be dealt with, simply:

(i) What is the impact of manufacturing processes as presently used and how may they be improved? Are there alternatives available or likely to become available?

1.2 Legislative Impact | 3

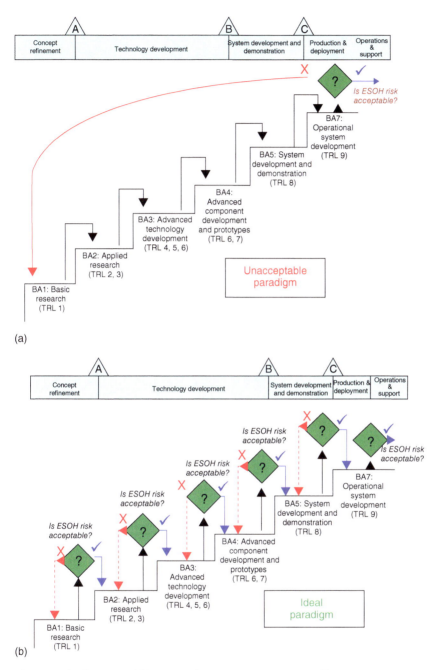

Figure 1.2 (a) Current and (b) proposed assessment practice – see Chapter 8.

(ii) What is the effect of use – on humans and on the environment?
   (a) What are the toxicity effects in handling and use?
   (b) What are the effects on land – that is managing contamination?
(iii) Are there disposal techniques available using safe methods?
(iv) Can improved disposal methods be devised?
(v) Finally, what are the costs involved? Are there spend-to-save options?

It is clear from examining the published literature that no one nation has all the answers and that no one nation has unique problems. While legal requirements do vary, there are common themes affecting all.

There is active work ongoing in the United States under the Strategic Environmental R&D Program (SERDP), a joint approach between Department of Defense (DoD), Department of Energy (DoE), and the EPA. There have been studies in the European Defence Agency and also studies in the North Atlantic Treaty Organization (NATO) – Science and Technology area.

These legislative requirements are driving research, as has been noted. However, they are discussed further in this book.

1.3 NATO Studies

Several activities have been completed or are in progress. Some have been openly reported [8, 9], but others may be available to NATO members and partners.

AVT 115: Environmental Impact of Munition and Propellant Disposal – study completed in 2009 and reported as an open document [10].
AVT 177: Symposium in Edinburgh 2011 – Munition and Propellant Disposal and its Impact on the Environment.
AVT 179: Design for Disposal of Present and Future Munitions and Application of Greener Munitions Technology (completed in 2013).
AVT 197: Munitions-Related Contamination – Source Characterization, Fate, and Transport (2012–2014).
AVT 269: Sea-Dumped Munitions and Environmental Risk (2016).

The first study, AVT 115, which was reviewed and discussed widely, produced the following conclusions:

- Open burning/open detonation (OB/OD) is not generally acceptable, although there are dissenting opinions and the use of amelioration technology is possible.
- Note that forensic studies have shown that residues do remain after detonation – these are used as court evidence. Whether these are meaningful in contamination terms needs discussion and examination.
- Technology exists for most current problems – current systems can generally be dealt with, although accidental failures or articles later discovered may need special treatment, and pyrotechnics can pose significant problems.
- Technology and needs are separated in many cases – e.g. the United States has technology/information and it is needed in, for example, Georgia.

- Availability of surplus systems must be considered as a target for terrorists as an easy source of materials.
- Surplus systems can also be targets for terrorist action, which may trigger an event.

There are therefore good safety and security reasons for dealing promptly with disposal.

1.4 New Ingredients and Compositions

It has been argued that changes in materials will answer the requirement and there is evidence that they can improve matters.

There is, however, a need to demonstrate that new materials offer significant advantages, and this is shown in several of the reports now in the open literature [5, 11–13]. An early example of this is the four-power programme on novel propellants [14]. Again, this is an illustration of the approach and, as detailed later, the area of focus is now materials such as ammonium dinitramide (ADN), etc (Figure 1.3) [15–21].

This was part of a multinational programme involving the United Kingdom, France, Germany, and the United States [14].

It involved joint studies on the formulation and testing of a smokeless propellant for tactical systems. The aim was proof of principle, but environmental issues did not play a major part in the study. It has interesting aspects, however, as elimination of acid smoke has been a first target for environmental improvement.

This is an improvement in many ways, but there are still products and these may be just as hazardous as the eliminated smoke. In some ways, an invisible product can be more hazardous.

Therefore, there is a need for clear demonstration of safety and proven ways of assessing true impact. This needs examination and experimental proof.

In short, simple answers can be in error and assumptions need testing before acceptance. These are the constraints that must be addressed.

There has also been considerable work on the replacement of metals in pyrotechnics and related systems [22–24]. The presence of metals, particularly Pb, is

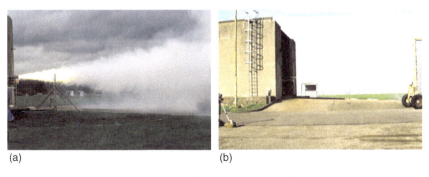

Figure 1.3 Comparison of (a) smoky and (b) smokeless propellants.

both undesirable and dangerous. Work has been under way for some time funded by the US Army with promising results [25]. Detailed toxicity studies are needed to avoid future problems of the kind found in the past and this is discussed in a later Chapters 8 and 10 in this book.

This is perhaps the most advanced study area, although small arms of all kinds are also being developed with the removal of ingredients of known toxicity. This is not as simple as might be supposed, as a recent Norwegian study [26] has shown. A round was introduced which seemed to offer improved environmental impact, but in use several Norwegian servicemen were taken ill prompting a detailed investigation. The results indicated that the new materials were less benign than originally thought. This illustrates the problems with the introduction of new materials where less is understood of their behaviour.

The development of national and international policies for the manufacture and use of less sensitive materials (insensitive munitions) led to the introduction and use of new polymer-bonded materials. While related to composite rocket propellants and themselves not possessing any significant problems, their manufacture make extensive use of isocyanates for curing the polymer. Many isocyanates are known carcinogens and therefore require careful handling, if not complete avoidance. To this can be added concerns over phthalates often used as plasticizers, which are now being banned in the EU.

Trinitrotoluene (TNT) has been used and is being used extensively and has been studied in depth by the US Army Corps of Engineers. It is toxic, but can be rendered non-available through immobilization in soils. It has useful explosive properties and ease of handing in preparation. This has prompted renewed research into similar materials to avoid some of the problems with polymer bonded explosive (PBX) while offering reduced sensitivity, and also to offer cost savings. However, recent studies have shown that it will leave more residues in use, and, more particularly, field disposal methods do not operate efficiently [27–29]. This is discussed in detail in later chapters.

1.5 Toxicology

It is hard to introduce new materials into use if there are uncertainties over their toxicity. Existing materials may well be toxic; but as the understanding of toxicity develops, their use may also be called into question [30–33]. For example, knowing how 1,3,5-trinitro-1,3,5-triazine (RDX) acts as a neurotoxin [31] helps manage the risk and should help devise treatment where possible.

This is a very active area and the likely main area of activity is in integrating this with other activities such as synthesis and formulation, as well as the study of the combustion and detonation products. It is often assumed that energetic materials are completely consumed when used in a design mode. However, forensic studies of explosives as detailed in the International Symposia on the Analysis and Detection of Explosives indicate that residues are left. An early paper [10] suggested that TNT could be trapped in explosives-generated carbon, for example. The question remains on the significance of those residues in health terms. Equally, the work by Walsh et al. indicates that significant residues are left by non-optimized function [30, 34, 35].

Contamination is of course not limited to the energetic materials, for metals in the system can be even more important and spread by explosive action [36].

First-generation tools now exist for modelling and predicting likely toxicity [37]. These have been developed for the speedier development of pharmaceuticals. These can be used to indicate bioactivity and hence estimate toxicity. Any such indication should reduce the testing and hence delays in introducing materials into use. It is important that they are used intelligently.

1.6 Life-Cycle Analysis

Environmental impact is part of the whole life of a munition and its ingredients. Experience elsewhere has shown that the whole life needs to be examined to understand and optimize the behaviour and so reduce the environmental impact. One of the areas identified for further immediate action within NATO was that of greener munitions. This formed the basis of a further study. Parts of the report are available and have been published [38].

At the outset of this study, the group identified several key issues that appeared to need examination:

Ingredients
Manufacturing
Use
Whole life-cycle management
Disposal
Impact on environment.

It became clear that the concept of greener munitions is far from simple. Not only are the individual aspects more complex but their interactions are also important and equally complex.

The approach and state of the art is discussed in later chapters.

1.7 Managing Contamination and Clean-Up

Land gets contaminated by use [2, 11, 14]. There is deposition from trials and tests as well as from impact and accidents. Often the use of ranges is poorly documented, and this is likely to be even more the case for battlefields. This is a prime source of contamination by hazardous materials, especially with incomplete functioning.

As reactive chemicals, energetic materials will have an effect on biology. This can be useful with nitrate esters being used to manage heart conditions, but on ranges, etc. it means that they can be bioavailable and therefore pose risks to health through incorporation into the food chain, perhaps through the water table.

For example, perchlorate is widely found in the water table, particularly in the United States, and as a bioactive material has provoked a series of programmes to understand its behaviour. Naturally, this has been extended to other energetic materials, with studies on behaviour and retention in soil and water. The behaviour depends on many factors including hydrogeology, soil structure, climate, and exposure. These all need consideration as do methods of assessing and managing any contamination.

Methods include bacteria and plants [39, 40] as well as more traditional chemical methods. Programmes on understanding the metabolism of energetics have been fairly successful and reported, with plants engineered to digest energetics. A problem arose in that energetic materials are not the preferred feedstock (other than ammonium nitrate) for bacteria; for example, energetic materials have less energy than more normal feedstock, although, of course, the energy that they have is released extremely rapidly in functional use.

As part of another multinational programme, there was a detailed study of ecotoxicology and land contamination. This work, involving the United Kingdom, United States, Canada, and Australia, was published in book form, but it forms a baseline for the assessment and management of land contaminated by energetic materials. It also includes a summation of the critical contamination levels as available at that point [35, 41, 42]. The first [41] report has produced a reference textbook on the Ecotoxicology of Explosives [43].

This publication [43] must mark the state of the art at the time, but requires updating on a regular basis to provide a measure of current understanding. However, the approach remains appropriate and the assessments and methods provide a sound basis for the necessary approach.

This is discussed in later Chapters 9 and 10.

1.8 Disposal Now and in the Future

The work by Walsh et al. indicates that there can be problems in disposing of new-generation materials as many of the existing tools for on-site disruption and disposal are insufficient for the task. This is specific for on-site disposal and would not affect the programmed demilitarization of surplus materials where a greater range of procedures can be employed. However, these are affected by legislation and the tightening of limits. The study in NATO indicated that most of the tools exist. These are being employed by various organizations to assist in the disposal of surplus materials worldwide. Some are being employed and developed by the NATO Support Agency under formal support agreements and are detailed in this book.

Unplanned disposal is not likely to diminish and cleaning up is certain to remain a live issue. The year 2018 also reminds us that material from the 1914 to 1918 war still requires handling!

These problems will continue and new variants will arise. The world situation means that tools for handling next-generation materials are needed, and tools must be applicable in a range of environments.

1.9 Recycling

Recycling is often seen as a way of covering the costs of disposal. However, experience has shown that at best it can be a disposal–cost offset. Metal parts can be recycled once certified free of explosives and the recovered energetics can possibly be reused for civil and military applications.

Techniques such as supercritical fluid extraction or liquid ammonia can produce recovered material which may be acceptable for use. However, a major drawback is the need to satisfy authorities of the consistency, and safety of the recovered materials. These materials need to be demonstrated to be safe in themselves and that no contaminants remain which will prevent safe use. This adds significantly to the cost. However, not all nations see this as an issue. It is likely to become more common especially with rare or expensive ingredients. It will require processes capable of producing a consistent product, or of making a consistent product from variable ingredients and hard evidence will be required to validate any such claims!

1.10 Conclusions

This is intended to provide an introduction to the technical area and to provide sufficient information to help manage environmental issues associated with munition systems.

In summary, to manage the potential environmental impact of energetic systems we need a range of approaches. Firstly, while it is not merely a matter of using new materials, they do offer sound options. However, they need to be understood well enough to deliver all the requirements placed upon them. This requires an understanding of likely toxicology and environmental and human impact as well as performance, ageing, and vulnerability. Since value for money also needs consideration, it may be that better specified and understood versions of existing materials will be more rapidly and effectively employed.

New processes can reduce manufacturing impact. Many processes were designed when there was less understanding of the effects and new approaches can be more efficient with reduced cost.

New-range management methods avoid damage and remove old damage. This is not limited to test ranges but also to manufacturing plants and storage facilities.

Overall, therefore, systems design for life minimizes overall impact!

These constraints and requirements should be considered a major driver for research and a scientific and engineering challenge. They require the following:

New methods for analysis.
New or re-engineered and well-characterized materials for use.
New methods for disposal.

References

1 Krause, H. and Fraunhofer ICT (1994). *Proceedings of the NATO ARW on Conversion Concepts for Commercial Applications and Disposal Technologies of Energetic Systems*. Moscow: Kluwer Academic Publishers.

2 Cumming, A.S., Mostak, P., and Volk, F. (2001). Influence of sampling methodology and natural attenuation of analysis of explosives content in contaminated sites. In: *7th International Symposium on the Analysis and Detection Explosives*. Edinburgh: DERA.
3 Branco, P.C., Schubert, H., and Campos, J. (eds.) (2001). *Defense Industries: Science and Technology Related to Security – Impact of Conventional Munitions on Environment and Population*, NATO Science Series IV: Earth and Environmental Sciences, vol. 44. Kluwer.
4 Cumming, A.S. and Paul, N.C. (1998). Environmental issues of energetic munitions: a UK perspective. *Waste Manage.* 17: 129.
5 Cumming, A.S. et al. (2009). Environmental management of energetic materials. *1st Korean International Symposium on High Energy Materials*, Incheon.
6 Ruppert, W.H. et al. (2010). A history of environmentally sustainable energetics. In: *Insensitive Munitions and Energetic Technology Symposium*. Munich, Germany.
7 European Commission – Environment Directorate General (2007), REACH in brief.
8 Cumming, A.S. (2008). Recent and current NATO RTO work on munitions disposal. *Proceedings of the 11th International Seminar on New Trends and Research in Energetic Materials*, Pardubice, Czech Republic.
9 Research and Technology Organization (2010). Environmental Impact of Munition and Propellant Disposal. *RTO Technical Rep. RTO-TR-AVT-115*, open publication. NATO.
10 Clench, M.R., Cumming, A.S., and Park, K.P.. (1983). The analysis of post-detonation carbon residues by mass spectrometry. *Proceedings of the International Symposium on the Analysis and Detection of Explosives*, Quantico, USA.
11 Jenkins, T.F., Grant, C.L., Walsh, M.E. et al. (1999). Coping with spatial heterogeneity effects on sampling and analysis at an HMX-contaminated antitank firing range. *Field Anal. Chem. Technol.* 3 (1): 19–28.
12 Betzler, F.M., Boller, R., Grossmann, A., and Klapötke, T.M. (2013). Novel insensitive energetic nitrogen-rich polymers based on tetrazoles. *Z. Naturforsch.* 68b: 714–718.
13 Talawar, M., Sivabalan, R., Mukundan, T. et al. (2009). Environmentally compatible next generation green energetic materials (GEMs). *J. Hazard. Mater.* 161: 589–607.
14 Cumming, A.S. et al. (2017). Performance tests of next generation solid missile propellants. *NDIA 2007 Insensitive Munitions and Energetic Materials Technology Symposium*, Miami.
15 Eldsäter, C., de Flon, J., Holmgren, E. et al. (2009). ADN prills: production, characterisation and formulation. *40th International Annual Conference of ICT*, Karlsruhe, Germany (23–26 June 2009).
16 Larsson, A. and Wingborg, N. (2011). Green propellants based on ammonium dinitramide (ADN). In: *Advances in Spacecraft Technologies* (ed. J. Hall), 139–156. IntechOpen 978-953-307-551-8.
17 Weiser, V., Eisenreich, N., Baier, A., and Eckl, W. (1999). Burning behaviour of ADN formulations. *Propellants Explos. Pyrotech.* 24: 163–167.

18 Weiser, V. (2017). Combustion behaviour of aluminium particles in ADN/GAP composite propellants. In: *Chemical Rocket Propulsion*, 253–270. Springer.
19 Rosenbaum, K., Bachmann, M., Gold, S. et al. (2008). USEtox – the UNEP-SETAC toxicity model: recommended characterization factors for human toxicity and freshwater ecotoxicity in life-cycle impact assessment. *Int. J. Life Cycle Assess.* 13: 532–546.
20 Anderson, S.P., am Ende, D.J., Salan, J.S., and Samuels, P. (2014). Preparation of an energetic-energetic cocrystal using resonant acoustic mixing. *Propellants Explos. Pyrotech.* 39: 637–640.
21 Coguill, S.L. (2009). Synthesis of Highly Loaded Gelled Propellants. *Tech. Rep.*, Resodyn Corporation, Butte, MT.
22 Steinhauser, G. and Klapötke, T. (2008). "Green" pyro-technics: a chemists' challenge. *Angew. Chem. Int. Ed.* 47: 3330–3347.
23 Brinck, T. (ed.) (2014). *Green Energetic Materials*. Wiley.
24 Sabatini, J.J. and Shaw, A.P. (2014). Advances Toward the Development of "Green" Pyrotechnics. In: *Green Energetic Materials*, 9-1–9-12. NATO.
25 Klapotke, T.M., Piercey, D.G., Stierstorfer, J., and Weyrauther, M. (2012). The synthesis and energetic properties of 5,7-dintrobenzo-1,2,3,4-tetrazine-1,3-dioxide (DNBTDO). *Propellants Explos. Pyrotech.* 38: 527–535.
26 Moxnes, J.F., Jensen, T.L., Smestad, E. et al. (2013). Lead free ammunition without toxic propellant gases. *Propellants Explos. Pyrotech.* 38: 255–260.
27 Walsh, M.R., Walsh, M.E., Taylor, S. et al. (2013). Characterization of PAX-21 insensitive munition detonation residues. *Propellants Explos. Pyrotech.* 38: 399–409.
28 Walsh, M.R., Walsh, M.E., Ramsey, C.A. et al. (2014). Energetic residues from the detonation of IMX-104 insensitive munitions. *Propellants Explos. Pyrotech.* 39: 243–250.
29 Walsh, M.R., Walsh, M.E., Poulin, I. et al. (2011). Energetic residues from the detonation of common US ordnance. *Int. J. Energetic Mater. Chem. Propul.* 10 (2): 169–186.
30 Lima, D.R., Bezerra, M.L., Neves, E.B., and Moreira, F.R. (2011). Impact of ammunition and military explosives on human health and the environment. *Rev. Environ. Health* 26 (2): 101–110.
31 Braga, M.F.M. et al. (2011). RDX induces seizures by binding to the GABAA receptor convulsant site and blocking GABAA receptor-mediated currents in the amygdala. *Proceedings of 'Human and Environmental Toxicology of Munitions-Related Compounds – from Cradle to Grave'*, Amsterdam, The Netherlands.
32 Perkins, E. (2011). Molecular toxicology of munitions in environmental species. *Proceedings of 'Human and Environmental Toxicology of Munitions-Related Compounds – from Cradle to Grave'*, Amsterdam, The Netherlands.
33 Meuken, B. (2011). Effects of combustion products of small caliber munitions. *Proceedings of 'Human and Environmental Toxicology of Munitions-Related Compounds – from Cradle to Grave'*, Amsterdam, The Netherlands.
34 Walsh et al. (2012). Munitions propellants residue deposition rates on military training ranges. *Propellants Explos. Pyrotech.* 37: 393–406.

35 Thiboutot, S., Characterisation of residues from the detonation of insensitive munitions, SERDP ER-2219. www.serdp-estcp.org/Program-Areas/Environmental-Restoration/Contaminants-on-Ranges/Characterizing-Fate-and-Transport/ER-2219/ER-2219.
36 Leffler, P., Berglind, R., Lewis, J., and Sjöström, J. (2011). Munition related metals – combined toxicity of lead, copper, antimony. *Proceedings of 'Human and Environmental Toxicology of Munitions-Related Compounds – from Cradle to Grave'*, Amsterdam, The Netherlands.
37 Schultz, T.W., Cronin, M.T.D., and Netzeva, T.I. (2003). The present status of QSAR in toxicology. *J. Mol. Struct. THEOCHEM* 622 (1–2): 23–38.
38 Ferreira, C., Ribeiro, J., Mendes, R., and Freire, F. (2013). Life cycle assessment of ammunition demilitarisation in a static kiln. *Propellants Explos. Pyrotech.* 38: 296–302.
39 Bruce, N.C. (2011). Plant and microbial transformations of explosives. *Proceedings of 'Human and Environmental Toxicology of Munitions-Related Compounds – from Cradle to Grave'*, Amsterdam, The Netherlands.
40 French, C.E., Binks, P.R., Bruce, N. et al. (1998). Biodegradation of explosives by the bacterium *Enerobacter Cloacae* PB2. In: *Proceedings of the 6th International Symposium on Analysis and Detection of Explosives*. Prague.
41 Brochu, S., Williams, L.R., Johnson, M.S. et al. (2013), Assessing the Potential Environmental and Human Health Consequences of Energetic Materials: A phased Approach. Final Rep., *TTCP WPN TP-4 CP 4-42, DRDC Valcartier SL 2013-626*.
42 Brousseau, P., Brochu, S., Brassard, M., et al. (2010), Revolutionary insensitive, green and healthier training technology with reduced adverse contamination (RIGHTTRAC) technology demonstrator program, *41st International Annual Conference of ICT*, Karlsruhe, Germany (June 29–July 02 2010).
43 Sunahara, G. (2009). *Ecotoxicology of Explosives*. Wiley.

2

General Introduction to Ammunition Demilitarization*

David Towndrow

General and Cooperative Services Programme, NATO Support and Procurement Agency (NSPA), 11 rue de la Gare, 8325 Capellen, Luxembourg

2.1 Part one – Logistics, Costs, and Management

2.1.1 Introduction

Following widespread reduction in operational ammunition stockpile requirements at the end of the Cold War, many nations were suddenly faced with large volumes of surplus munitions along with the need to make redundant many munitions production facilities. Around the same period, the previously accepted practice of deep sea dumping was outlawed and hence other disposal techniques were required. Investment by government and private contractors adapted a number of manufacturing facilities for the industrial demilitarization of large quantities of munitions. Industrial demilitarization is essentially the disassembly of munitions to separate the materials, recovering the commercially valuable materials and treating the hazardous materials to the point where they can be safely disposed of. In Europe, in particular, this is now a mature industry. From a Ministry of Defence (MoD) perspective, there is an imperative to deal with the full range of munitions, in various locations, quantities, and conditions; and this means having available a range of options to meet any one individual disposal/demilitarization action, including the ability to carry out open burning/open detonation (OB/OD) when required. Disposal/demilitarization must first be safe, then cost-effective, and environmentally responsible. Nations have different policies, largely driven by legacy capacity and interpretation of legislation and national policy. Most munitions are simple to demilitarize within the range of technical options currently available, with the final decision based on logistics, price, and availability of commercial facilities or the urgency of the disposal action. Some munitions are problematic, and this may increase the complexity of demilitarization and its costs, but the industry has proved to be good at resolving

---

* The views expressed in this report are those of the author and do not necessarily represent those of the NSPA or NATO.

Energetic Materials and Munitions: Life Cycle Management, Environmental Impact and Demilitarization, First Edition. Edited by Adam S. Cumming and Mark S. Johnson.
© 2019 Wiley-VCH Verlag GmbH & Co. KGaA. Published 2019 by Wiley-VCH Verlag GmbH & Co. KGaA.

potential problems and those involved in managing current munitions stockpiles should be confident that they will adapt for future requirements.

This chapter provides the reader with an overview of the issues faced by a nation in managing stocks of old or surplus munitions. It is written from a munitions management perspective that is broader than a simple list of demilitarization techniques and associated environmental and cost impacts. The first section describes the fundamentals of demilitarization with some specific examples. The second section provides more insight into the costs and environmental issues associated with disposal actions. Finally, there is a section on how nations might improve the situation, notably through making changes to the design of munitions to enable more efficient disposal. The work is not intended to address novel or exploratory demilitarization techniques in any detail. Developmental work is ongoing in industry, in academia, and in some nations through government-funded programmes. Some developmental work is publically reported and some only apply to very specific munitions and situations. Either way, the majority of this chapter focuses on the techniques that have served well, are currently available, and, through innovation, would likely provide nations with adequate demilitarization capacity into the future.

2.1.2 Context of Demilitarization

2.1.2.1 The Scale of the Issue

> *Of the hundreds of thousands of tonnes of munitions in national inventories over the last two decades, only a small proportion was used for its intended purpose, slightly more in training and the rest destined for disposal.................*

In 2018, some nations still have very significant surpluses of military munitions, much of it obsolete and ageing. Quantities vary depending on the historic national approach to defence and its associated stockpiling of munitions. An extreme example would be in Albania during the late 1990s with a surplus stockpile of over 200 000 tonnes of ageing munitions. Such surpluses require resources to monitor, store, and secure, and will present a risk of accidental explosion, sometimes on a catastrophic scale. Other nations have successfully driven down stocks of surplus munitions and implemented munitions procurement regimes to reduce the overall stocks and minimize any future surplus.

At the end of the Cold War in Europe, the authorities quickly recognized that surpluses would need to be disposed of, and that munitions production facilities were well placed to be adapted to undertake munitions demilitarization. Early investment by government and commercial operators to provide the capital equipment, particularly hazardous waste incinerators, allowed the development of a number of commercially viable munition demilitarization facilities. Today, there are a number of commercial demilitarization facilities across Europe bidding for contracts from various nations. Together, they provide a safe, effective, and environmentally responsible industrial solution for most, if not all, demilitarization requirements in the region.

Many nations have developed government/military-owned and/or operated demilitarization facilities, but these vary in scale and capacity from relatively simple low-volume operations to high-capacity full-range facilities. Similarly, some nations have a strong munitions research and manufacturing background, whilst others simply buy munitions and have simple munitions management regimes.

2.1.2.2 Factors Influencing Demilitarization

The type and quantity of ammunition in any one nation's inventory varies significantly. Some items are held in low quantity, others are continually used at training (or operations) and procured almost as a consumable with limited surplus for disposal, and others are stockpiled in high quantity (single purchase of whole of life stocks) with little ever used in training or operations with a potentially significant disposal requirement.

From a technical perspective, munitions items are generally grouped by characteristic, for example, mortar ammunition or air-dropped bombs, or by weapon effect, for example, smoke. From an operational perspective, there may be an important difference between two variants of small arms ammunition (SAA), but from a demilitarization perspective – assuming they are simply incinerated or pulled apart – they are substantially the same item. From the brief introduction given, it can be seen that it is important to consider all factors, not just the inherent munitions design, when considering any one decision on how to manage a demilitarization action.

National legislation and public acceptance will also have a major impact on a Defence Ministry's current and future ammunition disposal options. What is acceptable practice in one nation, or even in one location in a nation, may not be acceptable or sensible elsewhere: each case should be assessed separately.

Costs and associated logistics have a major influence on disposal actions. In Europe, transportation costs are typically between 20% and 30% of the total cost of the commercial demilitarization contract, i.e. to move, say, 500 tonnes of munitions from a military depot in Belgium to a commercial demilitarization facility in Italy. A nation may choose to store particular items, or consolidate with other nations until a bulk quantity is available in anticipation of lower commercial costs. For logistic resourcing reasons, a nation may choose to demilitarize items earlier than anticipated to allow for depot closure or simply freeing up storage capacity. Operational imperatives sometimes force non-optimal demilitarization decisions; for example, at the end of an overseas mission, a military may need to destroy damaged or limited quantities of munitions quickly and locally to avoid repackaging them prior to return to the home base. In small quantities, this could be accepted as an extension of explosive ordnance disposal (EOD) action that is used to destroy items judged unsafe for transportation.

Some nations have developed government-owned and government-operated sites, others use commercial contractors, and some have no demilitarization capacity, relying on uniformed personnel to destroy small quantities of munitions or arranging support from neighbouring countries. Most nations have the capacity to deal with EOD items under emergency arrangements. In North

Atlantic Treaty Organization (NATO), the NATO Support and Procurement Agency (NSPA) Ammunition Support Partnership provides a collective demilitarization service for NATO member and partner nations. In the period from 2010, NSPA has arranged demilitarization contracts for various nations at a total annual value of between 15 and 45 million euro/year.

Munitions are normally subject to a very high degree of security and accountability throughout ownership. Most governments strive to maintain accurate logistical data on the condition and location for all munitions under control. There is a high degree of confidence in the data and integrity of the logistic information systems in most NATO nations, but this is not necessarily the case in under-resourced nations or those still in the process of defence reform or recovering from major conflict. Often, there is limited technical information about the history of the munitions. Even with mature systems, sometimes things go wrong, such as mistakes in the delivery of the precise type, quantity, or packaging of munitions, mistakes in the provision of technical data, or even subtle changes in the design of munitions presenting problems for an optimized demilitarization process; for example, screws suddenly found to be 'glued in', or, more seriously, changes to explosive fillings in subcomponents.

Within a nation, the responsibility for different aspects of munitions management usually falls to different Defence Ministry staff branches. Overall responsibility for the through-life management of any one category or item of ammunition will usually rest with a nominated 'Munitions Manager'. That individual will coordinate all activity associated with the munition, from development, procurement, dispositions, and usage, through to repair/modifications and, finally, disposal. Some items require simple routine management, whilst others require a dedicated team. Other munitions-related activities tend to be centralized in other specialist departments. For example, the following activities usually fall to separate Defence Departments:

- specialist teams for concept and development of new munitions,
- storage in base depots or forward areas,
- transportation coordinated with other defence commodities, albeit moved as 'dangerous goods',
- the management of technical data,
- safety and performance tests and trials – including health and environmental assessments, and
- a munitions disposal team to coordinate the necessary activity between MoD departments and external industry.

Most nations will have developed clear policy and procedures to enable the different departments to work effectively together and provide advice, but it is the munitions manager who is ultimately responsible for coordinating activities for a particular munition and has responsibility for providing the demilitarization operators/contractors with critical information for their safety.

Ammunition demilitarization is a potentially hazardous activity, particularly where aged, damaged, or when out of anticipated specification ammunition is processed. Even under the most stringent regimes with the highest safety assessments, incidents still occur. Careful consideration is required when balancing the perceived environmental benefits and the risks associated with some demilitarization processes.

This chapter does not consider individual EOD procedures, sales, or transfers of munitions as a means of disposal. Neither does it cover deep-sea dumping or landfill/burial (a legacy activity now widely banned; and although technically sound for some items in some locations, it would be difficult to defend on safety or environmental grounds).

2.1.3 Demilitarization Process

2.1.3.1 Basic Stages of Demilitarization

A munition demilitarization and disposal process essentially dismantles the munition so that the different materials may be separated. Some items may have significant commercial scrap value, whilst others are hazardous and need further treatment or disposal as 'waste'. Whatever the process, it must be safe, economic, and environmentally responsible. Most processes follow the same basic process (Figure 2.1).

(i) *Removal from storage*: The demilitarization and disposal operation starts with collecting the munitions in suitable lots depending on the type and physical condition of munitions. The munitions must be labelled, controlled, and packaged as would be done for any other munition of its type. The munitions may then be transported and shipped to the organization or contractor responsible for the demilitarization and disposal operation.

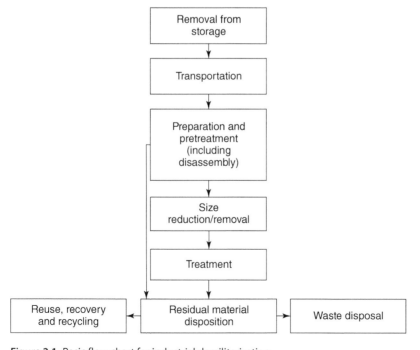

Figure 2.1 Basic flow chart for industrial demilitarization.

If munitions are found to be in an unsafe physical condition and cannot be transported, further inspection must be done to determine if EOD should be employed. The plan of action for a situation with munitions in an unsafe physical condition is beyond the typical scope of demilitarization, and a risk assessment may be needed to determine the appropriate emergency procedure required.

(ii) *Transportation*: Depending on the storage location of the munitions and the location of the demilitarization and disposal process site, various military or civilian regulations for transportation will need to be followed, especially if the transport involves crossing national or state borders.

(iii) *Preparation and pretreatment*: Munitions to be removed from military service use often involve a variety of materials, some of which do not present an explosive hazard, such as packaging materials and steel casings, and other materials, such as explosives and fuels, which are hazardous. After separation and screening, packaging materials, wood, paper, and metals should be collected for recycling, or disposal, according to the regulations for solid waste. Special attention must be paid to materials that require special treatment and disposal. The disassembly process will likely follow in reverse order the assembly procedures used in the production of the munitions. All hazardous materials should be identified for treatment by type. For example, igniters, fuzes, batteries, heavy metals such as lead, cadmium, mercury, asbestos-containing materials, will all be regulated at different levels depending on the final disposal process (Figure 2.2).

(iv) *Size reduction/removal*: The size and volume of a complete munition can usually be reduced by separating explosive warheads, rocket motors, and other large sections that contain hazardous materials by means of mechanical sectioning, laser grooving/cutting, water jet cutting, and cryofracture. Washout or melt-out processes are possible removal techniques. Whatever hazardous materials or hazardous components are remaining must be prepared and transported for treatment. The size reduction of explosive/pyrotechnic materials/components can lower the hazard from mass detonation to detonation of small parts or simply burning.

(v) *Treatment*: Any method or process designed to change the physical, chemical, or biological character or composition of any hazardous waste so as to neutralize such waste, or so as to recover energy or material resources from the waste, or so as to render such waste non-hazardous; less hazardous; or safer to transport, store, or dispose of.

(vi) *Reuse, recovery, and recycling*: Once the munitions have been separated from other inert materials, several options for recycling, reuse, and recovery of explosives, metals, and other materials exist. Options that provide the most advantageous cost benefit, such as recovery of explosives for industrial reuse, are selected. The demilitarization and disposal options resulting in the highest degree of reuse, recovery, and recycling of the most valuable materials will usually be preferred. However, care should be taken to not saturate a particular reuse market segment such that the recovered explosives become a logistical storage problem in their own right.

Figure 2.2 Typical process steps during industrial demilitarization. (a) Disassembly step for a sub-munition-filled artillery shell. *Source:* Courtesy of Nammo. (b) Saw cutting of HE-filled artillery shell. (c) Cut segments ready for melt-out of the explosive filling for recovery of energetic material. *Source:* Courtesy of D Towndrow. (d) Water jet cutting of artillery shell.

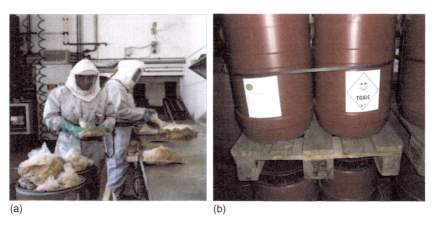

Figure 2.3 Recovering commercially viable materials and preparing any hazardous wastes of industrial demilitarization. (a) Recovered energetic material for civil reuse. *Source:* Courtesy of Spreewerk Luebben GmbH. (b) Ashes from an EWI packaged for approved hazardous waste disposal. *Source:* Courtesy of D Towndrow.

(vii) *Residual material disposition*: The disposal of small quantities of original demilitarized constituents that are a by-product of treatment (Figure 2.3).
(viii) *Waste disposal*: After completing the preceding steps, further processing may be required before the material is sufficiently safe to be released from government ownership. Options for demilitarization and disposal of waste military munitions and explosive hazardous waste are discussed later. Hazardous material refers to material that may pose a risk for the population, property, safety, or the environment owing to its chemical or physical properties or the reactions that it may cause. Authorized and trained personnel and permitted (and/or licensed) facilities will dispose of any materials remaining that cannot be recycled or reused. Inert substances usually can be disposed of at solid waste landfills, but hazardous materials must be disposed of at controlled and permitted facilities.

2.1.3.2 Demilitarization Facilities

All the demilitarization activities may be carried out at a single high-capacity site, or may be spread between sites. Typically, a site will be run as a single industrial complex, i.e. a discrete secure fence enclosing all the operational and administrative activities of a given facility. It is important to note that a site will normally demilitarize a range of munitions throughout the year, often from different customers and using the same processing equipment. That is, the incoming munitions may be clearly identified, but the scrap and waste materials may be an accumulation of several munition types and from different customers. Such sites would normally have one site-wide safety and environmental management system (EMS) covering all activities, supplemented by an accounting system to monitor the quantity/weight attributable to any one customer.

In some situations, the primary contractor may use subcontractors or perform partial demilitarization pending later operations or even return certain components to government. The scale and focus of an operation may vary; for example, investment may be made in dedicated high-value equipment for a certain task, such as the automated dismantling of a run of several millions of cluster munition bomblets or for the processing of high-hazard chemical munitions, or a site may simply engineer a minor modification to existing equipment to cater for the new process (Figure 2.4).

2.1.4 Demilitarization Techniques

2.1.4.1 Demilitarization Techniques and Processes

The physical destruction techniques available range from the relatively simple OB and OD techniques to highly sophisticated industrial processes. There are costs and benefits associated with each process. The most appropriate destruction technique in a given situation will depend primarily on:

(a) the resources available in the area,
(b) the physical condition of the ammunition, i.e. can it be moved,
(c) the quantity of ammunition and explosives in terms of economies of scale,

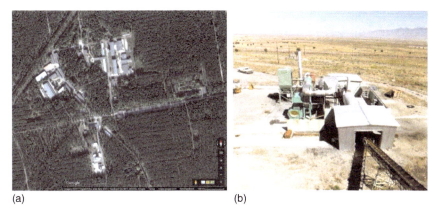

Figure 2.4 Industrial demilitarization facilities. (a) Satellite view of demilitarization facility in Europe. (b) The APE ammunition waste incinerator at Tooele Army Depot.

(d) national capacities and timeframe, and
(e) national policy regarding explosive safety, environmental legislation, and acceptability.

In general, the greater the amounts of ammunition to be destroyed, the larger are the economies of scale and therefore the wider range of affordable and efficient technologies (Figures 2.5–2.8).

2.1.4.2 Maturity and Use of Demilitarization Techniques

There are a number of discrete techniques and processes currently used for demilitarization. A summary of the most significant are given in Table 2.1. However, the more critical issue is the ability of a contractor to provide a demilitarization service to the required technical standards at an acceptable time and cost. A NATO Industrial Advisory Group[1] (NIAG) [1] study in 2010 provided a review of industrial capacity of the known major commercial providers in NATO and NATO partner nations, including their overall capacity and general capability. The report made a number of recommendations to increase efficiencies in the industry, but in general concluded that there was sufficient industrial capacity to meet both the types and quantities of surplus munitions. It noted that some nations still had significant surpluses but not necessarily close to the demilitarization facilities. Although the report is from 2010, the assessment of the industrial availability of the various techniques remains valid. The assessment only covered those methods that are either in use or likely to be widely used, and not small-scale or specialist techniques developed for specific munitions.

---

1 The NATO Industrial Advisory Group (NIAG) is a high-level consultative and advisory body of senior industrialists from NATO member countries, acting as a forum for free exchange of views on industrial, technical, economic, managerial, and other relevant aspects of the research, development, and production of defence and security equipment within the Alliance.

Figure 2.5 Open burning and open destruction. (a) Preparation for open detonation (OD) and open burning (OB). Whilst not an industrial demilitarization process, OB and OD are valid in some situations to minimize risks. (b) Open detonation in the United States. *Source:* Courtesy of D Towndrow.

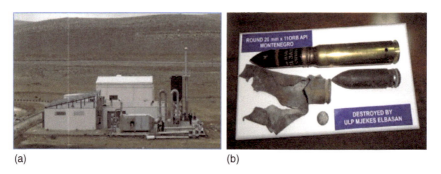

Figure 2.6 Small-scale incinerator and products of processing cannon ammunition. (a) Small-scale explosive waste incinerator. (b) Before and after incineration. *Source:* Courtesy of D Towndrow.

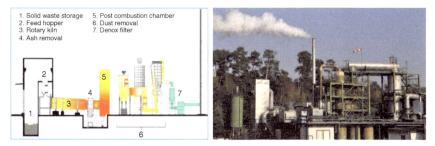

Figure 2.7 Large-scale explosive waste incinerator. Typical industrial hazardous material incinerator capable of destroying explosive materials, partially exposed explosives and for smaller items, the whole munition or warhead.

Figure 2.8 Automated, semi-manual, and manual processing. (a) High-capacity production line (melting out white phosphorous). (b) Labour-intensive production line (unpacking step of cannon ammunition demilitarization). (c) Remote automated handling equipment (high-volume high-hazard cluster munition process). (d) Semi-manual process (the worker is not present in the safety cell when the fuse is unscrewed). *Source:* Courtesy of D Towndrow.

The Organization for Security and Cooperation in Europe (OSCE) [2] published a Handbook of Best Practices on Conventional Ammunition in 2008. The section 'Best Practice Guide on the Destruction of Conventional Ammunition' provided a comprehensive review of all the techniques available, from OB/OD, through incineration and to alternate and experimental techniques. The handbook is due to be reviewed over 2018/2019; but as with the NIAG report, most of the widely used techniques remain the same.

Table 2.1 Generic list of demilitarization processes.

Technique	Description	Usage	Maturity	Notes
Disassembly and pretreatment				
Disassembly, manual	Disassembly, punching, crushing, unbolting, unscrewing or cutting by manual means	All munition types	Widespread use	Flexible – easy to adapt to different munitions; low setup costs; safety issues associated with manual; disassembly of munition
Disassembly, robotic	Disassembly, punching, crushing, unbolting, unscrewing or cutting remotely	All munition types	Widespread use	Reduces personnel exposure to munitions; less flexible than manual disassembly
Cutting, abrasive water or slurry jet	Sectioning by a high pressure abrasive water or slurry jet	All munition types	In use	Flexible and quick; generates waste water; useful with UXO and MEC
Laser grooving/cutting	Use of a laser to score a projectile case	All munition types	In development	Bisects the case to expose the filler
Cryofracture	Liquid nitrogen bath embrittles munitions before mechanical pressing	Small detonable items	Widespread use	Ensures no high order event during incineration; an additional process that may not be necessary
Removal				
Machining, mechanical	Dry machining of energetic materials by contour drilling	All munition types except small munitions	In use	No waste water; typically only removes 95% of explosive so further treatment is necessary
Washout, HP water	Ablation of energetic material by a high pressure water jet	All munition types except small munitions	Widespread use	Moderate pressures; generates waste water; difficult with small munitions
Meltout, steam	Meltout of explosive by steam or hot water jets	Melt-cast explosives	Widespread use	Generates waste water; moderate temperature and pressures
Meltout, autoclave	Meltout of explosive by hot water in a pressurized vessel	Melt-cast explosives	Widespread use	Reduced waste water generation; moderate temperature and pressure

Table 2.1 (Continued)

Technique	Description	Usage	Maturity	Notes
Destruction				
Open burning	Uncontained burning	Non-detonable items and waste	Widespread use	Simple process that does not require industrial plant; potential environmental impact
Open detonation	Uncontained detonation using a donor charge	Detonable items and waste	Widespread use	Simple process that does not require industrial plant; potential environmental impact
Closed detonation	Detonation by a donor charge in a contained chamber	Small detonable items	In use	Pollution control; transportable; small capacity batch process
Incineration, static kiln	Incineration in a sealed chamber. Items can burn or detonate	Munitions types except large detonable items after pretreatment	In use	Pollution control; small capacity batch process; items may require pretreatment such as sectioning before incineration
Incineration, rotary kiln	Incineration with the items slowly moved through the kiln	Non-detonable items and small detonable items	Widespread use	Pollution control; items may require pretreatment such as sectioning before incineration
Incineration, car bottom furnace	Incineration using a moveable 'car' to insert the waste	All munition types except large detonable items	Widespread use	May have pollution control; can handle unusual shapes; small capacity batch process
Incineration, plasma arc	Molten slag is heated by a plasma arc and destroys munitions	Explosive waste slurry or granular solids	In use	Can deal with pyrotechnics; slag encapsulates hazardous waste; items may require pretreatment such as sectioning before incineration; relatively expensive for bulk energetics
Biodegradation, aqueous/slurry	Biodegradation by microbes in a bioreactor	Explosive waste slurry or secondary waste stream	In use	In use for perchlorate treatment; cheap and environmentally acceptable option

(*Continued*)

Table 2.1 (Continued)

Technique	Description	Usage	Maturity	Notes
Reuse, recovery, and recycling				
Resale	Sale of munitions to foreign governments	Serviceable munitions	In use	Limited by arms proliferation agreements; munitions may be unsaleable by the time they are disposed of
Energy recovery	Co-firing of slurries or use of waste heat boilers on incinerators		In use	
Scrap metal recovery	Resale of scrap metal from demilitarized munitions		Widespread use	Requires flashing or some method of removing contamination
Chemical conversion	Conversion of energetic materials to saleable chemical products		In use	Typical applications include production of phosphoric acid from WP filled munitions
Energetics recovery	Solvent based techniques to recover energetics from cross-linked binders	Rocket motors and PBXs	In use	As a precursor to reuse as commercial or military explosives
Reuse as commercial explosive	Reuse of recovered energetics for commercial explosives	Recovered energetics	In use	

2.1.5 Summary

This section provided an overview of the techniques currently in use at commercial and governmental demilitarization facilities. It also provides an insight into the management issues associated with demilitarization such as cost, time, quantity, and specific requirements that in many cases actually drive the processes by which demilitarization is achieved.

2.2 Part Two – Environmental Issues and Demilitarization

2.2.1 Introduction

Industrial demilitarization is essentially the disassembly of munitions to separate the materials, recovering the commercially valuable materials and treating the hazardous materials to the point where they can be safely disposed of. In Europe, in particular, this has been a mature industry since the end of the Cold War in the early 1990s. Disposal/demilitarization must first be safe, then cost-effective, and environmentally responsible. Nations have different policies, largely driven by legacy capacity and interpretation of legislation and national policy. Most munitions are simple to demilitarize within the range of technical options currently available, with the final decision based on logistic, price, and availability of commercial facilities or the urgency of the disposal action. Some examples are provided to highlight the environmental and cost issues associated with specific demilitarization tasks. The second section deals with design for demilitarization (DFD), the process whereby final disposal/demilitarization is considered from the design development stage of a munition. Finally, a short section considers the future of demilitarization.

2.2.2 Demilitarization Process

2.2.2.1 Technical and Environmental Issues

Most demilitarization facilities, government or commercial, recognize the need to satisfy customers that national and international standards will be met and that all approvals are in place. Contractors often embrace international standards such as ISO 14001 for an EMS. ISO 14001 sets out the criteria for an EMS. It does not state requirements for environmental performance, but maps out a framework that a company or organization can follow to set up an effective EMS. ISO 14001 EMS would be certified by an independent awarding body and requires recertification every three years. The presence of such a recognized EMS indicates management attention to environmental issues, but not that any specific environmental impact is measured or is otherwise compliant with other standards. The EMS is normally linked to a safety management system and quality management system such as OHSAS 18001 and ISO 9001. The EMS should cover normal and abnormal impacts, and assess the risks and contingency plans associated with accidents/emergency situations. Some of the present and future constraints are discussed elsewhere in this book.

A facility will normally segregate any valuable scrap material and prepare it for sale to maximize the value, thus offsetting the cost of the process. Material such as brass from cartridge cases has a high commercial value, particularly if the metal is in a pure state and prepared ready for processing by a scrap dealer. Preparation usually involves a process of certification that the material is 'Free from Explosive Hazard' and sufficiently demilitarized as to prevent it being reused in a military role or being mistaken as a live item by a lay person (i.e. free from any live munitions, quantities of energetic material remaining in cavities or on surfaces, or other potential hazards but not necessarily completely free of traces of energetic material within the detection limits of, for example, airport security detectors). The material is often prepared in a form optimized for the next stage, for example, reduced in volume for transport to a smelting facility. Materials with little scrap value may be transferred to the local municipal waste, or recycling centres, and there are some examples of combustible materials being used as co-feed for the demilitarization site's own heating system. Hazardous waste is prepared for final disposal either directly to local approved landfill or is transferred for further processing at approved waste treatment facilities. The demilitarization facility will manage the flows of all the different materials to try to maximize the income from any commercially valuable material and reduce the costs of disposing of hazardous and non-hazardous waste. In Europe, in particular, there are comprehensive regulatory controls on the handling of 'waste' material. European Union Regulations apply directly to all member states' governments and come into effect on publication. European Union Directives are binding on member states that are then required to implement this in national legislation so as to meet the provisions of the directive by the required date. It is important to note that the national legislation may be applied in different ways in different nations. The local and national authorities are usually responsible for monitoring, compliance, and approval.

Most demilitarization facilities have an explosive waste incinerator (EWI) to incinerate the energetic materials. The larger EWIs are capable of incinerating complete items of ammunition such as SAA up to 30-mm cannon, or partially demilitarized munitions such as warheads, grenades, etc. There are a variety of EWI designs and methods of operation. The vast majority have an integrated flue gas treatment system to reduce emissions to authorized limits. The Industrial Emissions Directive (Directive 2010/75/EU of the European Parliament and of the Council of 24 November 2010 on industrial emissions [integrated pollution prevention and control]) is a European Union directive which commits European Union member states to control and reduce the impact of industrial emissions on the environment. It covers emissions from all operations from the site including the EWI. Note that there are minimum thresholds on the hourly/daily total quantity of material being processed for application of this directive (and national legislation), typically in quantities of tens of tonnes per day processing. In some circumstances, small installations for safely burning limited and occasional quantities of material may fall below the thresholds and the installation may be accepted as an 'oven' without pollution control equipment. Emissions to air are monitored, sometimes continuously, and form part of the regulatory oversight of the installation. In addition to monitoring normal operation, the systems will monitor any abnormal operation, i.e. on start-up/shutdown, and other occasions where higher than normal emission occurs.

Once transported for processing, most of the demilitarization activity will take place within one single facility. That facility will likely process different munitions, possibly from different customers throughout the year. It is possible to generate an 'environmental assessment' from the perspective of the munition undergoing demilitarization, i.e. to assess all the impacts from each stage of a munitions processing for a specific quantity, but such an assessment may not be particularly useful. If the purpose of such an assessment is, for example, to help inform a decision concerning open burning vs industrial processing of a large quantity of munitions over a sustained period, then it may have merit; but if such an assessment relates to munitions in small quantities, or for simple demilitarization processes, then the outcome of the assessment may be of questionable value compared with the effort of generating it. In the case of facilities with an EMS, the facility itself will be able to monitor and control the activities and associated environmental impacts from a site perspective and over a meaningful period, such as annually. Some estimates may be made should a particular customer require objective data, but, in general, an overall assessment of the efficiency and environmental impact of any one site should satisfy a customers' due diligence in selecting an appropriate contractor. There is, however, one important caveat, and that is that the facility must be able to process the munition, with all its specific hazards and materials, within the site's authorized discharges/environmental envelope (Figures 2.9–2.11).

Figure 2.11 provides an overview of the various waste streams from two demilitarization facilities within the same company. Nammo NAD in Norway is a deep mine specializing in 'closed' detonation where the ammunition is normally detonated without processing and only the packaging recovered, whilst Nammo Buck in Germany has a range of demilitarization operations, optimized to recover materials.

In summary, provided the facility has the appropriate techniques to dismantle and treat the components in an authorized way within the sites' approved EMS, there should be no unacceptable impact on the environment. The environmental discriminators between different MoD options for demilitarization then become energy use such as CO_2 release from transport, energy use in processing, etc. (Figure 2.12).

2.2.2.2 Open Burning (OB) and Open Detonation (OD)

OB/OD operations are conducted to destroy excess, obsolete, or unserviceable munitions and energetic materials such as explosives, propellants, and pyrotechnics (i.e. the demilitarization stockpile), as well as media contaminated with energetic materials (e.g. manufacturing wastes). OB/OD generally takes place on the ground surface or in earthen pits and trenches. Bermed areas, burn trays, and blast boxes can also be used to better control and contain the process and the resulting contaminants and emissions.

In many nations, OB/OD remains a common disposal method for energetic material, and in nearly all nations OB/OD are utilized by the militaries for small quantities or where there is no alternate disposal technique, such as items that are unsafe to move. This stems from its effectiveness, low cost, and capacity to treat and destroy a wide range of explosives, ordnance, and propellants, typically: bulk energetics such as blocks, pellets, powders, and liquids; energetic-contaminated wastes including gloves and plastic; energetic-contaminated containers such as wooden crates and cardboard boxes; small cased munitions like cartridge-actuated or propellant-actuated devices, fuses, small projectiles, and SAA; medium cased munitions including bomblets, warheads, rocket motors, projectiles, and missiles; and heavy cased munitions containing more

Nammo Buck GmbH
Emissionsdaten der Anlage zur thermischen Entsorgung von Explosivstoffen
für den Zeitraum 01.01.2015 bis 31.12.2015

Die Nammo Buck GmbH betreibt am Standort 16278 Pinnow eine Anlage zur thermischen Entsorgung von Explosivstoffen. Die Anlage ist nach der 17. Verordnung zur Durchführung des Bundes- Immissionsschutzgesetzes (17. BImSchV- Verordnung über die Verbrennung und Mitverbrennung von Abfällen) genehmigt. Die Anlage arbeitet seit 1993 im genehmigungskonformen Betrieb.
Die Emissionsmessungen zur Überprüfung der Einhaltung der Emissionsgrenzwerte wurden vom 12.10.2015 bis 15.10.2015 durchgeführt.

Allgemeine Daten		
Betriebsstunden im Berichtszeitraum	h	2930
Mittlere Rarchgasmenge	Nm³/h	4650

Kontinuierliche Überwachung (Tagesmittelwerte)		Jahresmittel	Grenzwert
CO	mg/Nm³	26,61	50
NO$_2$	mg/Nm³	85,65	200
SO$_2$	mg/Nm³	0,03	50
Staub	mg/Nm³	1,51	10
C$_{ges}$	mg/Nm³	0,70	10
HCL	mg/Nm³	0,02	10

Einzelmessungen bezogen auf 11 Vol-% O$_2$ (Mittelwerte)		Maximaler Messwert zuzüglich erweiterte Messunsicherheit ermittelt im Messzeitraum	Grenzwert
PCDD/PCDF	ng/Nm³	0,02	0,1
Hg	mg/Nm³	0,009	0,05
HF	mg/Nm³	< 0,29	1,0/4,0
Summe Cd + Tl	mg/Nm³	0,013	0,05
Summe Sb -Sn	mg/Nm³	0,09	0,5
Summe As + Cd +Co +Cr + Benzo-a-pyren	mg/Nm³	0,02	0,05

Erläuterungen:
h   Stunde
mg   Milligramm
ng   Nanogramm
Nm³   Normkubikmeter (Volumen eines Gases im Normzustand bei 273,15 K und 101,3 kPa)
Staub Gesamtstaub
CO   Kohlenmonoxid
NO$_x$ Stickoxide
SO$_2$ Schwefeldioxid
HCl   Chlorwasserstoff
HF   Fluorwasserstoff
PCDD/PCDF   Dioxine und Furane

Summe Sb- Sn:
C$_{gesamt}$ Gesamtkohlenstoff
Hg   Quecksilber
Cd   Cadmium
Tl   Thallium
Sb   Antimon
As   Arsen
Pb   Blei
Cr   Chrom
Co   Cobalt
Cu   Kupfer
Mn   Mangan
Ni   Nickel
V   Vanadium
Sn   Zinn

Die durch den Genehmigungsbescheid 049.00.00/92 festgelegten Emissionen wurden im Berichtsz-itraum sowohl während der kontinuierlichen Messungen als auch bei der Durchführung der Einzelmessungen eingehalten. Die Form der Veröffentlichung dieser informationen ist mit dem Landesamt für Umwelt abgestimmt.

Figure 2.9 An example of publically available emissions data as part of an industrial demilitarization facility in Europe (Germany). *Source:* Courtesy Nammo.

than 100 lb of energetic material per item (e.g. aircraft bombs, artillery shells). Although OB and OD are often referred to as one disposal process, they are individually quite different and involve different physical and chemical processes which impact the efficiency of the disposal and, often, the characteristics of contaminants emitted to the environment. Each process is explained in more detail here.

Environment

We will carry out our operations in such a manner that they cause the minimum amount of damage to the external environment.

Our ambition is to be a good neighbor. We want all operations within the industrial area to be environmentally friendly. We will satisfy the requirement in ISO 14001 and similar standards.

The annual audit at all sites ensures us that the focus on health, environment, safety, and security (HESS) is generally good. However, a few of our sites do have challenges regarding ground pollution and need to improve. For example, in Mesa and Vihtavuori the inherited pollution issues skill need to be resolved.

In 2015 the number of reported environmental incidents from our sites have increased.

The purpose of our risk analysis is to reveal and prevent factors that may threaten the environment. We continuously work to standardize our risk management system and increase internal training in conjunction with emergency plans.

We follow up and measure waste, effluents to water and emmisions to air, reduce noise levels and handle other environmental factors in accordance with existing regulations and internal instructions. It is important to react immediately to any deviation, inform the relevant inspection authorities and maintain an open-minded attitude towards employees and the public on environmental issues.

If an incident occurs, we need to ensure efficient protection measures are in place to avoid consequences for the environment. Further, it is essential not to use materials, chemicals or processes where hazards cannot be adequately controlled and to ensure that hazardous waste is handled in accordance with instructions.

Reducing energy consumption and good energy conservation measures have high priority at all our sites. Environmental considerations and cost reduction measures are important for our choice of energy sources. Since we started using natural energy (geothermal power), we have made significant energy cost savings. At the aerospace propulsion business unit at Nammo Raufoss, we invested two million Norwegian kroner when we installed a geothermal power plant. We have saved 65–70% on our energy consumption, when we compare geothermal energy to other traditional energy sources.

Figure 2.10 Example environmental statement from a commercial demilitarization group's website. *Source:* Courtesy Nammo.

Name of company	Country	Current year					
		Hazardous waste	Cardboard	Paper	Woodwork	Plastics	Residual waste
Nammo NAD	Norway	1 980	440	NR	12 080	NR	229 650
Nammo Buck	Germany	207 840	305 420	13 020	349 360	171 690	158 942

Figure 2.11 Example of publically available data from the annual EMS report of an industrial demilitarization facility in Europe (Germany) (Kg of waste). *Source:* Courtesy Nammo.

Figure 2.12 Mass detonation of ammunition in a deep mine facility in Norway. *Source:* Courtesy Nammo.

2.2.2.3 Open Burning

OB is the destruction of energetic materials by self-sustained combustion, ignited by an external source such as flame or heat. In some circumstances, dunnage (e.g. wood) and auxiliary fuels (e.g. fuel oil or kerosene) may be added to initiate or sustain the combustion process. Bulk propellants, explosives, and pyrotechnics that are not reliably detonable and/or can be burnt without causing an explosion are often treated by OB. In general, the following materials are best suited to OB:

- bulk propellants (either bulk or in cartridges or charges with combustible cases);
- SAA;
- pyrotechnics and simulators;
- rocket motors (either static fired or within cages); and
- small high-explosive (HE) charges such as detonators.

Secondary, detonating explosives can also be destroyed by burning when OD is undesirable (most usually to mitigate shock or noise), although a certain amount of preparation is usually required. On occasion, solvents that contain energetic constituents or other energetic-contaminated wastes are treated using OB. In the past, OB was regularly carried out on the ground surface or in burn trenches, although current best practice involves the use of burn pans or concrete pads to contain the energetic waste prior to treatment as well as residue and ash from the burn. However, OB can still take place directly on the ground when very large disposal items are involved, such as rocket or missile motors, where an explosive charge is used to break open the casing and the energetic motors are then allowed to burn, or for the burning of bulk propellant/propellant charges by laying out propellant trains (Figure 2.13).

A typical OB source will burn at approximately 538 °C, or 371 °C when burnt with dunnage. It is possible for OB to reach higher temperatures, for instance, in the thermal treatment of propellants, where temperatures can exceed 1272 °C for some materials. Another example includes the OB of UK Barmine, for which

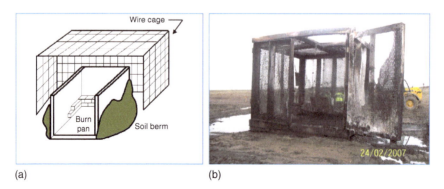

Figure 2.13 (a) Diagram of a burning cage and (b) photo of a partially damaged burning cage on a typical burning ground.

trials were carried out and the temperature of the burn recorded at 1620 °C. The following routine OB methods are used by a Western European military:

- OB of bulk propellant or propelling charges laid out as a propellant train on a wetted surface (normally bare earth);
- burning of SAA, detonators, pyrotechnics, primers, lachrymatory items, and augmenting cartridges within a gas-powered small arms ammunition destruction oven (SAADO), vented to atmosphere;
- OB of up to 60 kg of wetted gunpowder in steel trays;
- OB of smoke generators under specific environmental conditions (warm, dry, and windless day as far away from grazing land as possible);
- burning of pyrotechnics, primers, augmenting cartridges, and shotgun cartridges in burning tanks; and
- limited authorization to use modified burning tanks for the disposal of SAA on operations.

2.2.2.4 Open Detonation

OD is the destruction of energetic materials (i.e. detonable explosives and munitions) using a HE donor charge to initiate the items requiring detonation. Bulk explosives and whole-piece munitions and/or pyrotechnics are generally treated by OD. Due to the variety of explosives and munitions requiring disposal in this way, there is no typical OD event; however, the militaries of most nations have developed procedures that can be used to ensure the efficient destruction of items. Many of the procedures are relatively straightforward, although it is often the case that stacks of munitions requiring disposal consist of a variety of different types of munitions. Guidance is provided to assist the technician to lay out the stack in the manner most likely to result in the complete detonation of all of the items. There are a number of munition types that do not fit readily into the categories mentioned; and in such cases, specific-to-nature disposal procedures may need to be devised and authorized. Examples include the destruction of complex objects such as guided missiles, which may have a number of explosive components such as the warhead(s), rocket motor(s), safety and arming mechanisms, actuating charges, squibs, gas bottles, and flares, etc. A detailed understanding of the make-up of the item is required so that donor charges are best placed to achieve the most complete destruction. Other items, such as energetic-contaminated wastes, may require specific knowledge. Unlike the current best practice for OB, OD is typically carried out directly on the ground surface, in open pits or trenches, or via buried charges. The maximum temperatures associated with OD are, in general, significantly higher than those recorded during OB, ranging from 700 to 5600 °C. These temperatures are also greater than those required for hazardous waste incineration (which range from 500 to 1200 °C) OD involves a number of reactions and processes. These can be categorized as detonation, afterburning, air-entrainment, plume formation, and plume dispersion, and are outlined here.

Detonation The detonation reaction is extremely rapid and hot, reaching thousands of degrees (ranging from 700 to 5600 °C) in microseconds. Products of detonation include significant amounts of compounds, originating from the energetic materials being treated, that have not fully reacted. For example, there may be significant amounts of hydrogen, carbon monoxide, methane, ethane, and formaldehyde gases produced during the detonation phase.

Afterburning Afterburning immediately follows detonation and is visible as a fireball. The process requires an air atmosphere and involves the transformation of the incomplete reaction products (produced during detonation) to final, more stable products. Afterburning takes several seconds to complete, generally involves temperatures between 700 and 1700 °C, and produces additional CO_2 (from the reaction of carbon monoxide and air) and water vapour (from the oxidation of hydrogen gas). Intermediate hydrocarbon products will also be oxidized to form CO_2 and water. At this point in the OD process, metal fragments originating from the metal casings of the explosive materials being treated will have accelerated to high velocities, travelling outside and ahead of the fireball. Afterburning reactions are crucial in determining the composition and impact of final emissions to the environment from OD. Any practice that may alter or suppress afterburning (such as the placement of soil over the detonation pile before initiation in order to mitigate blast and noise) may result in incomplete combustion (i.e. where the more harmful intermediate reaction products are not converted to final, stable products) and an increase in risk to human health and the environment. The majority of the metals remains as solids and tend to remain in the immediate locality.

Air Entrainment Air entrainment involves the rapid expansion and mixing of detonation gases with the surrounding air; and as the incomplete gaseous products react with the air in the afterburning, the reaction volume continues to expand and rise. As the hot gases rise, they entrain additional air and this entrained air allows further combustion reactions to take place. Air entrainment and afterburning occur almost simultaneously and are highly coupled.

Plume Formation A highly visible OD plume is formed as dirt produced from the crater and surrounding surface soil is entrained along with the air and the rising gases. The time taken for a plume to form varies between seconds to hundreds of seconds.

Plume Dispersion The plume is clearly visible for many minutes after detonation has occurred. The plume will move downwind and, as it does so, it will continue to expand, further diluting the concentration of contaminants. Terrain will also influence the movement of the plume and the way in which contaminants are diluted and dispersed in the atmosphere.

Management Practices and Controls There are a number of management practices and controls normally in place in nations to regulate OB/OD. Although the majority of these are concerned with safety and security, there are ways in which the reported environmental risks and impacts from OB/OD may be managed and reduced through:

- optimizing the efficiency of the process by carefully planning and conducting OB/OD activities (e.g. ensuring that weather conditions, stacking techniques, burn temperatures, and source material mixes are optimal for reducing potential environmental impact);
- improving OB/OD facilities to include better containment, run-off control, and monitoring mechanisms (e.g. berms, ditches, covers, regular contamination analysis); and
- making use of alternative methods of disposal where possible.

National Perspectives   Some countries prohibit OB/OD activities as a means of logistic disposal entirely (notably Germany and the Netherlands) and others restrict them (e.g. in the United States and Canada). In the United States, permits or licenses are required before OB/OD activities can be carried out due to the safety hazards associated with the treatment of energetics, the potential for significant waste stream variability that may be difficult to predict and characterize, intermittent/quasi-instantaneous releases that are challenging to monitor and model, limited opportunities for engineering controls, and regulatory requirements for site-specific environmental performance standards. Therefore, to obtain an OB/OD permit in the United States, the following factors have to be considered and assessed: air emissions; casings, and other munition components; ejecta, unexploded ordnance hazards; soil explosives hazards; historical operations; and conceptual site models; whilst best management practices for OB/OD units are to be specified as permit conditions. No such system exists in the United Kingdom, where the use of OB/OD is generally permitted for the disposal of certain 'difficult' waste streams and, hence, is most notably used for the disposal of explosives or energetic materials. However, to ensure that OB/OD disposal is carried out in the most appropriate manner, in 2007 the Confederation of British Industry (CBI) [6] and the Health and Safety Executive (HSE) Explosives Industry Group, together with representatives from the MoD and others, jointly published a document entitled 'Guidance for the Safe Management of the Disposal of Explosives'. This document is being rewritten in 2019 and contains advice on the safety management measures that need to be employed for OB/OD, along with an overview of the pertinent health and safety legislation and the environmental controls required. The guidance document provides the United Kingdom with a national standard for OB/OD.

It would be irresponsible to suggest that OB and OD is an environmentally benign destruction methodology for all ammunition natures, yet substantial scientific research has taken place over the past two decades that suggests that the environmental impact of the destruction by OB/OD of some ammunition natures is relatively benign [5, 7, 8]. Similarly, preparation of OB/OD requires labour and other resources, and may require post-event clean-up, particularly from potential throw-out if the detonation is incomplete.

In summary, OB/OD remains an important disposal option for certain types of munitions, in carefully selected locations where the recovery of potentially valuable materials is not a major factor. As a technique, it will remain necessary in circumstances where there are no viable alternatives. Where there is choice, an informed and balanced view based on technical merits should be taken for each disposal action. Whatever the technical merits, future policy changes may affect acceptability and this needs to be part of any planning.

2.2.2.5 Examples of Cost and CO_2 in Demilitarization Options

Although some years ago, the United Kingdom's disposal of a large quantity of obsolete ammunition provided a well-documented example of the process a munition manager might follow to determine the 'best' option for any one disposal action. A significant effort was made to examine and then demonstrate that simply burning the items at a local approved site had, in overall terms, the least negative environmental impact. A balance had to be made between the resources used to generate the evidence to support the case, or spending the time and funds with industry to dispose of them. The case was regarded as important as it would provide supporting evidence for similar OB techniques to be used on similar items in the future. From an environmental perspective, it provided supporting evidence for the site's EMS. Prior to this work, there had been poor public perception of OB and OD activities at the site (albeit principally concerned with the noise of OD and other weapon testing activities), but a series of meetings with the local authorities and members of the public explained the proposed activity and was supported.

In 2006, UK MoD was considering options to dispose of around 200 000 surplus anti-tank 'Barmines'. The mines were old but in reasonable condition, each containing 8 kg of RDX/TNT. Sale to an approved third party was explored but discounted, so various options to demilitarize the items were considered. Although a large quantity of RDX/TNT was available in a relatively accessible form, the energetic material was of variable quality, and no commercial buyer for any quantity came forward at that time. Following a competitive process, the MoD identified two demilitarization options: one to have the mines defused in the United Kingdom and then shipped to a Nordic commercial demilitarization facility where the mines would be dismantled and the energetic material recovered for co-use in quarrying booster charges, and the second option was to burn the Barmines under controlled conditions at the UK-MoD-owned, QinetiQ-operated site on the Thames Estuary. Extensive work was undertaken to demonstrate that the emissions to air were sufficiently benign and would be sufficiently dispersed at the site's boundary (and other sensitive receptors), that the process would be acceptable to the local population and the regulator when compared to other options. As there was a small but credible risk of detonation of the mines during the burning process, a backup option was deemed necessary, and some 40 000 mines were sent for industrial demilitarization in parallel with the main burning programme. The 40 000 represented the minimum quantity to justify the establishment of the demilitarization line and associated transportation, partially by ship. The slides summarize the work and the outcome. The carbon from the energetic materials themselves would ultimately be released to the environment whichever option was chosen and so was not a discriminator (Figure 2.14).

Disposal of 170 000 Barmines (8 Kg RDX/TNT in plastic case plus a fuse)

- Considered: Reuse, recovery of RDX/TNT, use in training/Ops
- Developed two options – burning in UK £1 and commercial processing in Europe £13.
- Burning this quantity would be highly contentious. Significant trials programme to determine atmospheric and terrestrial releases. Full site specific sustainability appraisal. Best practicable environmental option. Independent review by Manchester University. Ministerial approval. Brief MPs and local authorities. Burnt over 100 000 with no complaints.

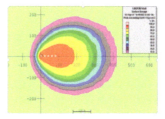

The work to determine the environmental impact

- Over 90 000 previously disposed by open burning. Previous work demonstrated environmental and safety acceptability but work not considered good enough to support a formal review in 2007.
- Theoretical study to determine products of combustion suggested acceptability.
- Two series of burning trials to determine quantitative levels of potential contaminants and other aspects particularly the dispersal and visual impact of black smoke.
- QinetiQ and Dstl joint work to model the fate of contaminants and assess the risks to human health and the environment under a range of metrological conditions.
- Site specific sustainability appraisal.
- Defence estates regulatory review.
- Dstl review of the body of evidence
- Manchester University independent review against industry best practice.

Two options considered in detail – open burning and processing at a commercial facility in Sweden

A review of the overall cost and CO_2 emitted (from the energy used in transportation and processing) for the two options provides a clear differentiator, noting that 1190 tonnes of CO_2 is emitted from the materials in the Barmine whichever option is chosen:

- The commercial processing would take place at a commercial facility in Sweden and cost £2.25 M. CO_2 generated from the transport alone for this option is calculated at just over 450 tonnes.

- The disposal of Barmine by open burning at QinetiQ Shoeburyness would cost around £250 K and associated transport would produce around 4.3 tonnes of CO_2.

Figures 2.14 Example of cost and CO_2 used to examine two options for a major disposal.

A further example relating to the transportation of large quantities of Multiple Launch Rocket Systems (MLRS) rocket pods from the United Kingdom and Western Germany is given here to demonstrate the impact on overall CO_2 emissions when the transportation phase is included in the scope of the environmental assessment. The 42 truck movements involved in that specific example (total 80 tonnes CO_2) is relatively small when compared to other routine daily MoD logistic truck movements, or, for that matter, the daily truck movements by any mid-sized commercial logistics organization. For comparison, flying Frankfurt to Philadelphia in a modern efficient airliner is quoted by Lufthansa as 0.9 tonnes CO_2 per passenger for the round trip, or for a full flight with 300 tonnes of CO_2. Effectively, the CO_2 budget for the whole transport task was less than 50% of one return aircraft flight from Germany to the United States (Figures 2.15 and 2.16).

One final example (see Figure 2.17) is the preliminary planning that was undertaken by an international organization considering options to support the Albanian MoD in 2005. The aim was to provide an estimate and compare

⌂ Your flight:
From: Frankfurt (DE), FRA to: Philadelphia (US), PHL, Roundtrip, Economy Class, ca. 12,700 km, 1 traveler

CO_2 amount: 0.949 t

Figure 2.15 CO_2 emissions for a typical long-haul flight.

Transportation: cost and CO_2

400 MLRS pods (42 trucks) (2013)

Total demil cost 2.8MEUR incl Ia. Transport options were:

1. UK Depot to Eastern Germany 1400 KEUR (multimode) 80 ton CO_2

2. Western German Depot to Eastern German facility 90 KEUR (truck only) 60 Ton CO_2

Transport costs typically one third of total demil cost: CO_2 very significant – NO_x, noise etc?

Air cargo - 0.8063 kg of CO_2 per ton-mile
Truck - 0.1693 kg of CO_2 per ton-mile
Train - 0.1048 kg of CO_2 per ton-mile
Sea freight - 0.0403 kg of CO_2 per ton-mile
Source: OECD

Figure 2.16 Typical CO_2 estimates used to examine two options for the transportation element of a major disposal. *Source:* OECD.

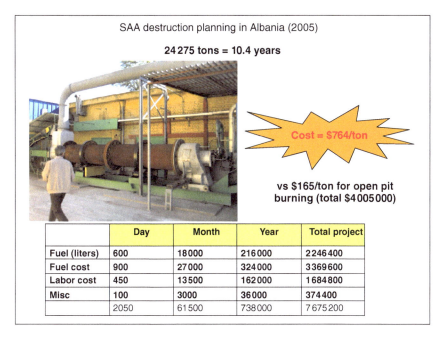

Figure 2.17 Example of a preliminary estimate to determine open burning vs EWI incineration of SAA.

options of burning a very significant quantity of SAA (steel cased with limited commercial value for the scrap) compared to incineration through an EWI. Clearly, there were issues of potential ground contamination and air emissions associated with the proposal to burn around 24 000 tonnes of SAA, but these may have been manageable and have been acceptable to the national authorities in the remote location under consideration. The additional CO_2 emissions of 2.8 Kg CO_2 per litre fuel oil led to a daily emission (from 600 l of fuel oil) of 1.68 tonnes of CO_2. For various reasons, including donor acceptability, the EWI option was chosen. This example is provided to highlight the cost and CO_2 emissions when proposing the EWI. It should be noted that some nations prohibit such open burning, some organizations prohibit the practice within any commercial industrial demilitarization contracts, and in all cases the other potential negative impacts must be considered. Environmental assessments should be balanced (or focused), proportionately and only used where there is a specific reason to do so.

There are a number of mobile or transportable demilitarization plants commercially available and in use. They range from the relatively simple single-truck/containerized systems to modular systems with multiple containerized units linked to provide higher capacity or handle different munitions/techniques. Such systems can be moved to the ammunition site, thus significantly reducing the transportation element. See an example in Figure 2.18.

 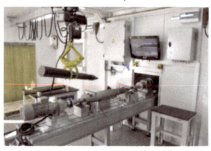

Mobile demilitarisation plants – environmental impact

ISO Cnr based EWI incl pollution control 5 tonnes 5.56 mm SAA in 8 hrs

EODSolutions' Transportable Ammunition Destruction System (1-RADS).
© EODSolutions/2012

Modular ISO Cnr based units designed for specific process lines

The ammunition disassembly component of JAKUSZ's Planetarium transportable system, processing a 100 mm UBK8 projectile. © JAKUSZ SZB/2011

Figure 2.18 Examples of mobile demilitarization plants.

2.2.3 Design for Demilitarization (DFD)

2.2.3.1 NATO AOP 4518 (Revised 2018)

Munitions designers have traditionally focused their product development and design on performance, not on ease of demilitarization at the end of the munitions' life cycle. DFD/disposal was not a significant issue when less complex disposal techniques, e.g. OB/OD, or dumping were utilized.

Because legislation is likely to become more stringent in the future, plans and processes used in the demilitarization and disposal of munitions will need to be carefully crafted and be reviewed continually in the light of new legislation and advances in technology. The focus of demilitarization and disposal is escalating from using techniques that are safe, efficient, and cost-effective to ones that are environmentally acceptable, physically safe, free of health hazards, practical, and cost-effective. Nations are being urged to move from disposal processes that rely on destruction towards those that maximize the recovery and reuse of component materials. Consequently, demilitarization has moved towards recovery, reuse, recycling, with demilitarization requirements of conventional ammunition (including any item containing propellants, explosives, or pyrotechnics) being considered early during munitions system design.

In NATO, there are a series of working groups charged with coordinating policy on munitions safety and associated testing regimes. They arrange the drafting and publication of NATO Standardization Agreements (STANAG) [3] and Allied Operating Procedures (AOP).

Safe disposal of munitions, design principles and requirements, and safety assessment (originally STANAG 4518, Edition 1 dated 8 October 2001) has recently been entirely rewritten (Edition 2 dated 24 May 2018). The guidance is aimed at all those with a role in munitions management, but specifically those engaged in the design of new munitions and their associated disposal techniques. It is available in the public area of the NATO Standardization Organization website and provides a useful guide to for nations when developing new munitions. The following is a summary of the AOP.

The agreement is applicable to new developments; existing stores subject to major modification, rework, change, or addition of any hazardous component; replenishment purchases; and existing stores being used in a new role. Participating nations agree to procure for military use by NATO Forces munitions purchased off the shelf which have been or will be modified, to comply with AOP-4518. The nation developing a munition shall implement the design and assessment principles detailed in AOP-4518. Specifically, the nation developing a munition agrees to incorporate acceptable end-of-mission (EOM), end-of-operational-life (EOOL), or end-of-life (EOL) disposal capabilities; to assess the design for its adherence to the guidance provided here: and to document the EOM/EOL disposal processes to be used on the munition.

New techniques have been developed to increase the use and benefits of reuse, recovery, and recycling (R3) methods. R3 methods break down the munitions into their basic, recyclable component parts and compounds, which can then be sold to offset processing costs and thus reduce the overall cost of demilitarization. However, not all munitions are suited to R3, and in some cases insufficient quantities exist to develop economically sustainable demilitarization solutions and operations.

The efficiency of these demilitarization operations is, among other factors such as the quantity and rate of demilitarization, significantly influenced by the original design of the ammunition. Yet, the design of munitions does not always lend itself to the cost-effective recovery of materials for recycling or reuse. Munitions designed and purchased today tend to be more complex than that currently being demilitarized. Munitions may incorporate embedded electronics, plastic-bonded explosives (PBX), and insensitive high explosives. These munitions designs may not efficiently accommodate current demilitarization processes, presenting additional difficulties during disassembly, and ultimately adding time and expense to demilitarization operations (Figure 2.19).

DFD is the incorporation of demilitarization consideration throughout the life cycle. The objective is to produce munitions that can be demilitarized safely and more efficiently at a lower cost. This involves incorporating features into the munitions design with a systems engineering approach that facilitates demilitarization processing. Desired outcomes include ensuring that:

- components can be disassembled easily.
- energetic materials can be removed.
- demilitarization processes can be used efficiently.
- munitions are safe to handle by operators throughout demilitarization processing.
- reusable or recyclable components/materials can be economically recovered.
- there is minimal environmental impact.
- ultimately, cost savings to future development programmes may possibly be realized through the use of recovered components and materials in new production.

Figure 2.19 Practical demilitarization steps. (a) Initial disassembly of Multiple Launched Rocket System (MLRS). (b) Some stages are automated, some semi-manual. Decisions are based on the quantity, throughput, and availability of equipment/facilities as much as the design of the munitions, although if certain DFD features are present, clearly it may assist. *Source:* Courtesy of Nammo.

The following design safety principles should be applied during munitions development to facilitate demilitarization and disposal using processes that maximize safety and minimize health hazards, negative environmental impacts, and life-cycle cost:

- Select materials that are not inherently toxic and can either be reused, recycled, or destroyed with minimum impact on health and the environment at the end of the munitions life.
- Select materials and design features that will minimize the adverse impact of credible service-life environments and aging on demilitarization and disposal processes and by-products.
- Select materials and design features that allow old operable stocks to be consumed in training.
- Configure munitions for safe disassembly and ease of useful material recovery.
- Configure munitions for ease of component and package reuse or recycling.
- Design munitions to maximize service life.
- Design munitions to permit significant life extension modifications and, consequently, reduce the need for demilitarization and disposal.
- Design for ease of alternative munitions applications with limited remanufacturing.
- Select materials and design features that allow the munitions to continue to be safe for handling, transportation, and storage even after end-of-life.

Personnel for new and modified munitions development programmes shall prepare an appropriate demilitarization and disposal plan. Uses of the related documents listed in AOP-4518 are recommended for consideration, when applicable, during the development of a plan. In general, a demilitarization and disposal plan should include the following information:

- A functional and physical description of the munitions (including quantity), its packaging configuration, and the equipment, processes, and procedures planned for safe and environmentally acceptable demilitarization and disposal;
- A listing of all materials, including the hazardous materials contained in the munitions, and their associated hazards;
- An indication of intent to conduct a hazards analysis on the demilitarization and disposal procedures and EOL actions in accordance with AOP-15 and to include a discussion of the safety and environmental impacts and their associated hazards;
- Provisions to ensure that, after application of the selected processes, all sensitive materials and items will be neutralized or otherwise rendered inoffensive or be extracted for other uses;
- The intended destination of liberated hazardous materials.

The AOP requires an assessment of the proposed techniques, accepting the limitations of such a contemporaneous review for an event potentially two decades in the future. The assessment should be to the Best Available Techniques

Not Entailing Excessive Costs (BATNEEC) principle. There may be more than one technique to demilitarize similar munitions, and the focus of demilitarization should be according to the BATNEEC principle, where:

- 'Best' means most effective in achieving a high level of physical and environmental protection.
- 'Technique' includes both the technology used and the way in which the installation is designed, built, maintained, operated, and decommissioned.
- 'Available Technique' means those developed on a scale which allows economically and technically viable implementation.
- 'Not Entailing Excessive Costs' implies where the benefits gained are worth more than the costs of obtaining those benefits.

Whilst it may take many years to renew all items in a national stockpile, the introduction of munitions that are easily disposed of with minimal environmental impact will allow significant savings over the life cycle of the inventory.

2.2.3.2 A Munition Manager's Perspective of Disposal Plans

From a munition manager's perspective – the individual or organization responsible for taking decisions, as set out in Section 1.3 – the degree of detail of the disposal plan or design for disposal effort, will be strongly influenced by the following factors:

- The type of munition and its inherent design and materials – a simple design, few hazardous materials and with current, mature demilitarization options may not need any specific plan or assessment.
- Large munitions, with potentially high quantities of energetic materials, would likely need specific disassembly procedures and possibly associated rigs and tools.
- Munitions containing specific hazardous materials or other hazards that might be encountered during disassembly where a disassembly procedure would be beneficial to a future contractor.
- Munitions held in large quantities and not used in training that may require a high degree of effort for their planned disposal at end of life.
- As a minimum, EOD procedures for use by the Armed Forces for the emergency destruction of the item, either when it is unsafe to move, or for the disposal of small quantities of munitions when alternate industrial demilitarization is not available for whatever reason.

AOP 4518 provides a basis for considering disposal issues as part of whole life assessment, i.e. Design for Demilitarization. Munition disposal plans must be produced at an early stage in the procurement cycle. However, any plan needs to recognize that:

- It is the demilitarization contractor's capability that largely determines the specific demilitarization process.
- The disposal will be managed against legislation and capability at that time – i.e. there is little point in assessing a demilitarization action in detail against 2018 legislation when disposal will occur in 2028 or later.

- Disposal plans need to be proportionate and appropriate – in some cases, they may not be required at all.
- An MoD must sustain safe, environmentally responsible, and practical options for the disposal of a wide variety of munitions that meet logistic and cost constraints, including the need for OB/OD in some situations.

2.2.3.3 Future Trends

There are a number of centres exploring potential demilitarization techniques such as bio-remediation, photo-catalytic degradation, supercritical water oxidation (SCWO), or molten salt oxidation; but these tend to deal with the energetic materials and, other than in some specific circumstances, may not be practicable as a commercial option for the complete item of ammunition. They may well be useful for dealing with some of the traditionally problematic materials such as polymer-bonded energetics in large rocket motors, but the challenge is usually the physical separation of the energetics which can then be incinerated. Industry is familiar with the processes to break the ammunition down and can be expected to seek greater efficiency through innovation and increased automation, whilst at the same time reducing the risks of an accident and improving the environmental efficiency. Such developments are incremental and not generally reported outside the Industry. There are however a few publications and a well-known symposium that provide an insight into current and future trends:

The NIAG report published in 2010 provides a useful overview of the techniques, their availability and technical readiness level.

The NATO Funded Munitions Safety Information and Analysis Centre (MSIAC) [4] published a comprehensive review of demilitarization technology in 2006.

The U.S. Department of Defense (DoD) continues to host a bi-annual Global Demilitarization Symposium and Exhibition, with the 2017 event having been held in Lancaster, Pennsylvania. The presentations of the Symposia hosted by the U.S. NDIA can be found at https://ndia.dtic.mil/ under archive.

The symposium continues to support the DoD in global efforts directed at reducing the stockpile of excess and obsolete strategic, tactical, and conventional munitions. The symposium features Product Director for Demilitarization Updates; Headquarters/Program Reviews and exhibits. The agenda focuses on ongoing demilitarization/disposal, resource recovery, recycling, and reuse operations and programs, sale of recovered demilitarization materials, demilitarization R&D efforts, transitioning technologies; and environmental, safety and policy issues that affect the business. Contributions are normally invited from MoDs and industry both from the United States and eligible nations. The presentations from previous symposia are available on the web. The 2017 focus was 'Solutions to Munitions Demilitarization/Disposal Challenges':

- Safety/environmental-related initiatives
- Technology developments/emerging technologies
- Innovative use and sale of reclaimed/recycled material
- Component reuse initiatives
- Design for demil.

References

1 NATO (2010) Study on NATO Industrial Capability for Demilitarisation and Disposal of Munitions. *Final Rep. of NIAG SG.139, PFP(NIAG)D(2010)0010* (not publically available). NATO NIAG.
2 Organization for Security and Co-operation in Europe (2008) OSCE Handbook of Best Practices on Conventional Ammunition. https://www.osce.org/fsc/33371?download=true (accessed 15 June 2018).
3 STANAG 4518, Edition 2 & AOP 4518 Edition 2 (2018). http://nso.nato.int/nso/zPublic/ap/PROM/AOP-4518%20EDA%20V1%20E.pdf and http://nso.nato.int/nso/zPublic/stanags/CURRENT/4518EFed02.pdf (accessed 15 June 2018).
4 MSIAC (2006). *Review of Demilitarisation and Disposal Techniques for Munitions and Related Materials*. NATO. http://rasrinitiative.org/pdfs/MSIAC-2006.pdf (accessed 15 June 2018).
5 SEESAC (2004). SALW ammunition destruction – environmental releases from open burning (OB) and open detonation (OD) events. http://www.seesac.org/f/tmp/files/publication/304.pdf (accessed 15 June 2018).
6 UK Confederation of British Industry Explosives Industry Group (2018). *Guidance for the Safe Management of the Disposal of Explosives*. Confederation of British Industry http://www.eig2.org.uk/wp-content/uploads/disposal_guide.pdf (accessed 15 June 2018).
7 Erickson, E., Chafin, A., Boggs, T. et al. (2005). Emissions from the Energetic Component of Energetic Wastes During Treatment by Open Detonation. *NAWCWD TP 8603*, Naval Air Warfare Centre Weapons Division, China Lake, CA.
8 Boggs, T., Atienza Moore, T., Heimdahl, O. et al. (2004). Metals Emissions from the Open Detonation Treatment of Energetic Wastes. *NAWCWD TP 8528*, Naval Air Warfare Centre Weapons Division, China Lake, CA.

3

Assessment and Sustainment of the Environmental Health of Military Live-fire Training Ranges

Sonia Thiboutot and Sylvie Brochu

Defence Research and Development Canada – Valcartier Research Center, 2459 de la Bravoure road, QC, G3J 1X7, Canada

3.1 Introduction

The importance of understanding and minimizing the environmental impacts of human activities has grown over recent decades. The exponential growth of the worldwide population stresses the need for sustainable human activities. In recent years, the Armed Forces around the world have realized that they must also ensure the sustainability of their properties. Military live-fire training activities are an essential activity to ensure the readiness of our troops; and, therefore, live-firing ranges represent major assets for the Armed Forces which must remain open and sustainable. As the extensive use of munitions might generate source zones of munitions constituents (MCs) in the environment, it might also threaten range sustainment. Following the discovery of adverse environmental impacts in Canadian live-fire training ranges [1–6], a huge programme was dedicated to extensive range and training area (RTA) characterization; and it was further conducted in close collaboration between Defence Research and Development Canada (DRDC) and the Engineer Research and Development Center (ERDC)–Cold Regions Research and Engineering Laboratory (CRREL) in the United States [7–14]. The fate and transport of these emerging contaminants were studied by various laboratories, both in Canada and in the United States, in collaboration with DRDC and CRREL. This work was supported by many stakeholders in Canada and in the United States including DRDC internal funding, Department of National Defence Director General Environment (DGE), Director Land Environment (DLE), and the US Peer-Reviewed Funding Program, Strategic Environmental Research and Development Program (SERDP) (https://www.serdp-estcp.org/). The topic of heavy weapons range sustainability was recently integrated within the Nordic cooperation, a major multilateral agreement between Denmark, Norway, Sweden, Finland, and Iceland. A specific subgroup was created, in which Finland, Norway, Denmark, the United States, and Canada were involved, and was referred to as the Environmental Protection for Heavy Weapons Ranges (EPHW), with the objective of producing a final report

---

Energetic Materials and Munitions: Life Cycle Management, Environmental Impact and Demilitarization,
First Edition. Edited by Adam S. Cumming and Mark S. Johnson.
© 2019 Wiley-VCH Verlag GmbH & Co. KGaA. Published 2019 by Wiley-VCH Verlag GmbH & Co. KGaA.

in 2018. The Finnish Defence Administration has integrated the topic of range sustainability within the European Conference on Defence and the Environment (ECDE), held every two years (http://ecde.info/). ECDE is supported by an expert-level group, including environmental focal points and specialists from the Ministries of Defence of Member States, and Nordic Defence Estates. Live-firing range sustainability also attracted a rising interest from many North Atlantic Treaty Organization (NATO) nations, and various NATO Applied Vehicle Technology (AVT) task groups and specialists meetings were held to further define the extent of the problem and promote the advancement of knowledge in this area [15–19]. Note that references [16, 20–22] are not fully available, as they are restricted to NATO nations and Partners for Peace, while they were still referenced for the information of the readers on their mandates, objectives, and topics presented or discussed. A comprehensive report on munitions-related contamination was produced by the NATO AVT-197 task group, followed by a NATO specialist meeting in November 2015. This specialist meeting attracted more than 75 participants from 16 countries, highlighting the high priority of range sustainability across participating nations [18, 19]. In the light of the strong interest generated and the high level of priority of range sustainability, the NATO Collaboration Support Office tasked AVT-197 to develop a NATO Cooperative Demonstration of Technology (CDT) on military range characterization. The NATO AVT-249 task group was created with this objective in 2015, and the CDT was successfully held in the United Kingdom, in October 2016. The CDT attracted 31 participants from 16 countries, and six subject matter expert instructors transferred their expertise to the attendants. Based on the success and interest generated by the 2016's CDT, another one was held in the fall of 2018, again in the United Kingdom [20]. The interest in RTA assessment and sustainment of military operations is therefore very high in many NATO nations. This book responds to this high interest, and this chapter covers the context in which the awareness towards RTA sustainability arose and gives an overview of how to define the source of MCs, their associated risks, and how to mitigate them.

3.2 Background and Context

Even quite recently, it was believed that the use of munitions would lead to the dispersion of forensic traces of residues at the detonation point. This is true for most of the high-order detonations and, consequently, there was no awareness that RTA might be at risk due to a munition's environmental footprint. DRDC Valcartier was a pioneer in this area, when a team sampled live-firing ranges looking for munitions' residues. Various samples came back with concentrations of concern for MCs, which led to the creation of an internal Research and Development (R&D) programme dedicated to the understanding of the problems and identifying potential solutions [1–6], and a first draft of an assessment protocol was published in 1997 [7]. In particular, surface soils collected in an anti-tank range impact area were found to contain quite high levels of octahydro-1,3,5,7-tetranitro-1,3,5,7-tetrazocine (HMX) in the surface soils [8]. It was the first time that a risk related to the dispersion of MCs in military live-firing ranges was identified. The paradigm that only traces of MCs were dispersed in live-fire training was proved wrong, and the lack of knowledge appeared as a threat to the sustainability of RTAs. In the United States at that time, most of the efforts were dedicated to the measurement of MCs in explosives and propellants production sites, where important contamination arose from old batch processes.

As US scientists had developed robust analytical methods and tools for MC sampling in production sites, a meeting was held between US and Canadian experts to present the Canadian findings related to RTAs. This drove a very strong interest of scientists from the United States and the first joint paper was published in 1999 [9]. This topic was given very high priority by the chain of command of both countries, which triggered a multi-year collaboration between Canada and the United States to better understand the extent and nature of MCs in North American RTAs [10–14]. In the same period, the strategic importance of range sustainability was further proved when, in the United States, a major RTA was closed in 1999, due to groundwater plume contamination by 1,3,5-trinitrohexahydro-s-triazine (RDX), perchlorate, and lead. The US Environmental Protection Agency (EPA) discovered RDX in that single source of drinking water, a supply to more than 520 000 people of the town of Cape May in New Jersey. Directly above this aquifer was the 14 000-acre target area of a vast firing range where Army National Guard troops have trained for decades by launching ordnance and munitions, testing their power and precision on the training fields of the Massachusetts Military Reservation. This triggered the closing of the range, which was thus lost for training, and to a multi-year decontamination process with over 350 million US$ spent in 2010, and still ongoing today [23]. This crisis made the Canadian–United States collaboration on range sustainability highly visible, supporting the building of a large R&D programme. A very strong stakeholder network from Canada and the United States supported this and invested millions of dollars to better understand the dispersion of MCs and their fates. It was clearly demonstrated that the use of live ammunition and explosives inherently results in the deposition of heavy metals and energetic materials (EMs) into soil, surface water, and ground water. Depending on the nature and magnitude of such deposition, the introduction of MCs into RTAs may pose a health risk to the military users and/or to the surrounding population; impact local flora and fauna; and/or result in non-compliance with environmental legislation and policy. The risks associated with MCs threaten the ability of RTAs to remain suitable for and capable of supporting assigned institutional and operational training requirements. The awareness related to this risk also rose in other NATO nations, and joint efforts through NATO AVT led to the sharing of knowledge, to the holding of specialists meetings, workshops, to the publication of detailed NATO reports, and to NATO CDT. The ability to accurately assess the condition of our ranges and to take corrective actions when appropriate will enable the implementation of high-priority programmes, such as the NATO Connected Forces Initiative, which will result in an increased tempo and size of multinational live-fire training to better integrate NATO military forces.

The challenges encountered in assessing and managing military live-firing ranges are also expected for legacy sites, or ranges that were closed, and which now need to be assessed and cleared for alternative uses and even for battlefield sites. The assessment process for these ranges is complex and involves many steps, beginning with the collection of detailed historical data. The next biggest challenge for legacy sites is the presence of unexploded ordnance (UXO), both on the surface and beneath, which represents a high risk when alternative land uses are proposed. While the presence of UXOs is expected and tolerated in active ranges, their clean-up is mandatory in legacy sites, in order to allow other land uses. In Canadian RTAs, periodic sweeps are performed for surface UXO neutralization, but most of those that are buried are never found and never actively sought. Identifying and remediating UXOs is phenomenally expensive, since no technology presently exists that can reliably and cost effectively locate them without a great deal of human labour. These extensive efforts combined with the security requirements account for the high cost of clearing them. Another significant challenge is the fact that many of them are buried deep underground, representing a challenge from both detection and removal standpoints. Legacy sites may also include chemical warfare UXOs, which represents another high level of risk. Because of all these facts, it is almost impossible to reach a 100% removal of UXO clean-up, and a certain level of risk will always remain. Depending on the future use of the legacy range, the characterization of the surface soils and water bodies might also need to be undertaken. The same protocols as described for active ranges must be applied, with the caveat that the analytical parameters shall be broadened to include legacy munitions' analytes. The fate and transport of legacy MCs also need to be considered, to forecast in which environmental matrices they might have ended. With the growing of the worldwide population and the corresponding encroachment on military properties, it is expected that the pressure for assessing and cleaning these ranges will grow. On top of UXO clean-up, remedial actions might be needed if soil or water contaminants are detected. Some of the clean-up methods developed for active ranges could be applied, while some others might be developed for specific legacy MCs.

The challenges associated with legacy sites also apply to the assessment and management of battlefields, where actual war was conducted. This would include recent conflicts such as in Iraq and Afghanistan, as well as historical conflicts, such as World War I (WW I) and World War II (WW II), where problems emerged through poor understanding and management of problems. It is estimated that millions of unexploded munitions are still to be found in farmlands around battlefields. The question of who shall pay for the assessment and management of war zones is a sensitive issue that needs to be addressed when conflict ends. Another level of complexity is specific to a war zone, as it might involve a wide array of munitions from various countries in impact areas, as well as minefields. In recent years, asymmetric conflicts involve conduct of intensive wars in urban scenarios, impacting actual cities. In these cases, active live-firing military range protocols would not always apply. The assessment and management of former war zones might use some of the protocols developed for active ranges, while they may differ profoundly and drive the need for specific approaches. In war zones where many NATO nations are involved, any potential future assessment and clean-up might be shared in a multinational effort.

3.3 Munition-Related Contaminants

The detection of the type and quantity of contamination from MCs and their breakdown products in water, soil, and sediments from RTAs is vital for assessing the extent of contamination and ultimately assessing the risk to human and ecological receptors. The contaminants that might be dispersed in the environment following live-fire training are mainly energetic materials (EMs) and metals. Conventional weapons for live-fire training use propellant and main explosive charges. While a high-order detonation of a round that functioned properly leads to the deposition of traces of residues, there are many events that might take place and lead to higher levels of residues, either at the firing position or at the detonation point. After having invested years in the characterization of major North American RTAs, we now have a better understanding of the mechanisms that lead to the deposition munition-related contaminants in concentrations of concern and where the problematic areas exist.

The explosive formulations used in most of the conventional munitions stockpile are 2,4,6-trinitrotoluene (TNT), Composition B (TNT/RDX) and octol (TNT/HMX), or similar materials. Consequently, the explosive analytes mostly detected in RTAs are RDX, TNT and its main degradation products 2-amino-4,6-dinitrotoluene (2-A-DNT) and 4-amino-2,6-dinitrotoluene (4-A-DNT), and HMX [9, 14]. The propellants include both rocket and gun propellants. Many rocket propellants consist of a rubbery binder filled with ammonium perchlorate (AP) oxidizer and may contain powdered aluminium as fuel. The propellants can also be based on nitrate esters, usually nitroglycerine (NG), nitrocellulose (NC), or a nitramine such as RDX or HMX. Gun propellants are usually referred to as single base (composed of NC), double base (NC and NG), or triple base (NC, NG, and nitroguanidine [NQ]). A single-base propellant may contain 2,4-dinitrotoluene (2,4-DNT) with traces of 2,6-dinitrotoluene (2,6-DNT). Most gun propellant formulations are either single or double base and include metallic lead, which represents another inorganic environmental contaminant. The propellant formulations contain other minor ingredients such as stabilizers, plasticizers, and burn rate modifiers, but they represent less than 2% by weight and have never been detected in soil surfaces. The propellant-related MCs mostly detected in RTAs are 2,4-DNT, 2,6-DNT, NG, and perchlorates (ClO_4) [13, 14]. Table 3.1 presents MCs mostly detected in RTAs and the media in which they tend to accumulate. NC is not included in this list, as it is a polymer, considered non-toxic and non-bioavailable.

The usage of munitions also leads to the accumulation of inorganic contaminants, deposited by a variety of processes. High-order detonations disperse fine metal particles, except for pre-fragmented rounds, which produce large fragments. Much larger fragments are generally produced by low-order detonations. Metals can be transformed into other compounds, not originally present in the munitions. This transformation occurs during the detonation process or during

Table 3.1 Munitions constituents (MC) mostly detected in RTAs.

MC	Surface soils	Surface water	Groundwater
RDX and HMX	✓	✓	✓
TNT	✓		
2-A-DNT and 4-A-DNT	✓		
2,4-DNT and 2,6-DNT	✓		✓
NG	✓		
NQ	✓		
ClO_4		✓	✓

weathering of the metallic particles deposited on ranges. During a detonation event the temperatures and pressures reach extremely high values, which might exceed the melting temperatures of some metallic compounds. These molten species are free to react with other compounds to form new alloys, metallic complexes, or salts, which will all have their own environmental fate. After dispersion in the surface soil, both chemical and physical weathering of the metallic particles will take place. A whole suit of metals can be dispersed in RTAs and, therefore, the best choice is to use an analytical method which allows the widest possible range of metal analytes. A very interesting study was published in Finland on the fate of metals in the context of old munition stockpile open detonation. They conducted that study at the largest Finnish open detonation site in the region of Lapland, where most of their obsolete ammunition has been demilitarized since 1988. Using sand traps and spruce needles, they proved that when open detonating munitions in a sandy environment, most of the resulting metallic species were silica-metalloid salts, non-soluble and non-available [24]. In fact, in their study, the open detonation led to the vitrification of the metals, which is considered a remedial action. This would happen in specific conditions, so it is still critical to assess the levels of metals in RTAs, especially in ranges such as small arms ranges. The metals mostly encountered in RTAs are lead, copper, zinc, and aluminium, while a whole suit of metal analytes have also been reported.

3.4 Surface Soil Characterization in Live-fire Training Ranges

Range assessment represents many challenges, the first one being high resistance to any action from range managers, as they perceive it as a risk to their training programme, or worse, that it might lead to range shut down. It is vital to explain that the assessment team is there to help them understand their range's environmental health, allowing range managers to take any action needed to avoid potential range closure. The benefits of range assessment are numerous: it is a strong proof of due diligence; if a problem is identified, timely corrective actions can be applied to eliminate or mitigate the problem; it improves the health and safety of the military users; and it ensures the health and safety of the surrounding communities (often composed of military families). As soon as this is understood, the range managers and operators will normally support the range assessment team in achieving their goal. Another source of strong resistance within defence departments is related to investing in R&D to define characterization methods and tools for measuring MC footprints in RTAs. The argument was that industry would be able to perform these tasks, as they normally do for other types of contaminants, and that the industry would do it better than governmental R&D entities. In fact, at that time, the industry had never extracted, tested, or analysed EMs from environmental matrices and everything had to be developed through R&D. In Canada and the United States, major funding was dedicated to developing the required new tools and methods. The level of complexity was very high, as a result of considering the numerous MCs used in conventional ammunition, as well as the need for an intimate knowledge of the numerous types of live-fire training in a multitude of environments. In the United States, the transfer of expertise for sample treatment and processing to the industry was successfully done recently, while we still struggle to succeed in getting representative results from the industry in Canada.

3.4.1 Safety Aspects

Sampling in RTAs represents a high safety risk, as they are heavily impacted with UXOs. All the personnel involved in the sampling campaign need a mandatory safety briefing given by the range control office. When entering a UXO-contaminated range, the sampling team must always be accompanied by an experienced explosive ordnance disposal (EOD) specialist, who will indicate the safe path for walking and driving. In very high-density UXO areas or in antitank impact areas where piezoelectric fuses might be easily triggered, access might be either denied or restricted to the EOD specialist, who could perform the sampling after being given precise sampling instructions. This aspect is even more critical when performing groundwater surveys in RTAs.

3.4.2 Data Quality and Sampling Objectives

Soil sampling for MCs follows the same principles as for any other type of contamination. It involves systematic planning steps, depending on the sampling objectives. Samples are typically taken from defined areas within the area of interest. The areas defined as the 'area of interest' are established through a site planning process. These areas are typically referred to as sampling areas or decision units (DUs). Sampling of these areas is based on methods such as the data quality objective (DQO) process, which establishes the statistical limits to be applied to the analytical data obtained from the samples. The quantity of samples to be collected is determined on the basis of site usage and a general risk assessment (DQOs). So, the area where MCs might have been dispersed is defined as a DU, where the sampling will be conducted. A DU can be defined as an area where a decision is to be made regarding the extent and magnitude of contamination with respect to the potential environmental or human health hazards posed by the exposure to munitions contaminants. A lot of knowledge and skill has been developed on how to determine the size and location of the DU, and on the methods to collect soil samples. Mostly, the areas of interest are within the impact areas, where explosive residues are to be monitored, and firing positions where propellant residues are to be expected. The primary objectives in RTA characterization are as follows:

To measure surface soil concentration of MCs that may pose a threat to the health of military users and may reach the contaminants (human exposure);
To measure the surface soil concentration of MCs that may further be dissolved and brought to surface water bodies (human and ecological exposure);
To measure the surface soil concentration of MCs that may further be dissolved and reach the groundwater (human exposure);
To measure the surface soil concentration of MCs that may pose a threat to local ecological receptors (ecological exposure);
A combination of some or all these.

It is a huge challenge to obtain representative results of the level of contamination by MCs in RTAs that cover many square kilometers and where a multitude of activities involving munitions are conducted. The nature of explosive and propellant dispersion comprises both compositional and distributional heterogeneity. The compositional heterogeneity is due to the nature of the explosives and propellants: the formulations are complex and are inhomogeneous in nature from their conception. In other words, compositional heterogeneity is described as the variability of contaminant concentrations between the particles that make the population, which leads to a fundamental error. The fundamental error is managed by collecting and analysing sufficient sample mass to address the compositional heterogeneity. There is also a high distributional heterogeneity in the dispersion of MCs. Solid particles may vary from very fine dust to large chunks of explosives, up to centimetre size. This heterogeneity results in a segregation error. To minimize the segregation error and compensate for the distributional heterogeneity, multiple subsamples must be collected. To achieve that, it is strongly recommended to use a multi-composite sampling strategy with a systematic judgemental random sampling design to characterize the average concentration of MCs within a chosen area or DU. Multi-increment sampling is used to provide reproducible estimates of mean contaminant concentrations within a sampling area. A minimum of 100 subsamples is recommended to form a sample that weighs between 1 and 2 kg [18]. The size and location of the DUs are planned using knowledge sufficient to delineate areas that are likely to be contaminated. The DU is walked in a serpentine matter and increments are collected at each three to four steps, in the same area of the subunit. When the entire surface of the DU is covered, the same process is repeated after a rotation of the sampling path by 90°. The recommended sampling pattern is illustrated in Figure 3.1, where the red and blue arrows show the two perpendicular sampling paths.

Using multi-composite sampling, reliable estimates of mean concentrations for the specified area of virtually any size are obtained. Replicate samples are taken to determine if DQOs have been met. Specifically, replicates demonstrate whether the samples taken are reproducible or not. Replicate samples are not the

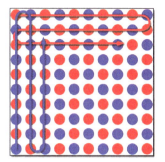

Figure 3.1 Surface soil sampling pattern.

same as field splitting of a sample, as a replicate sample is a unique sample within a DU that is taken in the same manner as the original one. Replicates should be taken on a regular basis during sampling, and the number of replicates in a specific DU should be at least three to allow a minimal statistical analysis. The number of replicates and the number of DUs requiring replicate sampling is a function of the DQOs and relies on the judgement of the leading environmental officer and will lead to defensible analytical results. Properly collected replicate samples should lead to a relative standard deviation lower than 30% between replicates. This is the only method which leads to a reliable estimate of the mass of contaminants in the DU area and can be further used to take informed decisions to ensure sustainability.

Except for ranges where the surface is regularly physically moved, the highest concentrations of MCs are always present in the top 2.5 cm surface soil. A variety of sampling tools are available to collect soil samples. Representative soil samples can best be obtained using a core-sampling tool. Indeed, CRREL has developed a coring device that is handy and most helpful in RTAs. It involves a soil plunger which can be adapted to vary the sampling depth, and the corer diameter can also be varied depending on the sampling goals [18]. The choice of collection tools often depends on the cohesiveness, coarseness, and moisture content of the soil. Scoops and spoons are necessary for non-cohesive soils and heavily cobbled surfaces. Coring tools are recommended for cohesive surface soils with and without vegetation, but they should be used whenever possible. Samples should be stored in a polyethylene bag, tightly closed. Splitting the sample in the field to reduce the volume sent for laboratory analysis must not be done, as the representativeness will be lost. The whole samples must be stored in the cold and dark and sent to the laboratory for homogenization and analysis.

This approach is different from the collection of discrete samples and the commonly used practice of field splitting or laboratory subsampling by removing only a portion of the sample received from the field for further processing. Implementing a multi-incremental sampling approach poses a challenge as there is a very high resistance in many nations to accept its application in RTA characterization. The unfortunate assumption is that it leads to sample dilution and that discrete sampling is more appropriate. That needs to be countered using reference sources, such as the NATO AVT-197 report, or NATO CDT course material, where more information on sampling strategy and tools can be found [18, 20].

3.4.3 Importance of Soil Sample Processing to Ensure Representativeness

The heterogeneous nature of MC mandates that care be taken with careful sample homogenization as samples taken from DUs will have contaminants heterogeneously distributed within it. Therefore, sample processing in the laboratory is as important as the sampling itself, as it ensures whole sample representativeness. Air-drying of the soil samples must be conducted in the dark and at room temperature, to avoid the alteration of the contaminants of concern. Once the samples are dried, they need to be homogenized to ensure that the subsample is representative of the whole sample. Then, prior to analysis, the contaminant must be extracted from the soil matrix. Over the years, various means of whole sample homogenization have been studied. In Canada, a method involving a slurry for the whole sample with acetone and rigorous mixing gave good results, although it was not applicable in all cases, such as when samples contained high concentrations of NC [7]. Only two methods lead to representative extracts: whole sample extraction with acetone or subsampling of a properly ground sample. Slurry extraction requires large amounts of solvent, but it does not require any specialized equipment. Extracts for analysis are typically filtered prior to injection in analytical instrumentation.

However, when sampling for MC, we have proved that a large soil sample is critical and therefore the whole sampling extraction is not the recommended method, as it would produce a lot of waste solvents and non-practical extractions. Instead, the recommended method is to subsample prior to extraction; this requires grinding of the whole sample to reduce the particle size of the matrix and contaminant and allow the mixing of the contaminant uniformly throughout the matrix. The recommended method for sample homogenization relies on mechanical grinding to reduce the size of the EM particles and fibres to the same size as the soil particles to ensure uniform dispersion. A properly ground sample significantly reduces the error associated with subsampling, allowing reproducible subsamples to be taken. Sample grinding requires specialized equipment, but generates less hazardous waste. After thorough homogenization, the whole sample is spread out on a clean surface and a subsample is built from at least 30 subsamples collected across the sample surface, using a systematic multi-incremental approach as done in the field, and is extracted with acetonitrile and analysed following the EPA 8330b method [25]. It must be stressed here that there is a strong resistance in commercial laboratories to running this protocol. The resistance generally comes from the fact that most laboratories do not possess any grinding equipment; moreover, they see this approach as expensive and time-consuming. The practice of grinding soils to ensure subsample representativeness is well known in the mining industry, as important investments depend on the reliability of the sample analytical results. Unfortunately, whole sample homogenization is not commonly conducted in most commercial environmental laboratories for several heterogeneous analytes, leading to non-representative results in many instances.

In Canada, round-robin tests were conducted to validate the precision and accuracy of extraction and analysis results obtained by commercial laboratories for MC-containing samples that were built by spiking soil and water with known quantities of ground energetic materials. It was demonstrated that, while water samples led to representative results, soil samples led to non-representative results, despite precise instruction on how to handle them. This highlights the importance of performing a thorough laboratory quality assurance/quality control (QA/QC) by running some spiked samples to assess the external process and to convince laboratories to adopt the processing method to generate reliable soil concentrations.

Therefore, when sampling in RTAs for MC, it must be stressed that samples for analysis should never be subsampled by scooping a small mass from an unprocessed sample. Samples should never be field-split or split prior to grinding. A processing request should be included in the sample shipment, specifying the type of processing needed to ensure error reduction in the subsampling process.

When metals are the analytes of interest, it is also essential to process the soil samples before analysis. The purpose of sample treatment is to produce a smaller, drier, and manageable sample suitable for laboratory-scale analysis while at the same time ensuring that the prepared sample is homogeneous and representative of the original field material. First, the samples should be dried at temperatures below 40 °C to avoid the potential loss of volatile compounds, such as antimony, arsenic, and mercury, and to avoid the oxidation of some metals. The whole sample should then be sieved using screens to remove pebbles, sticks, and fragments larger than 2 mm, and then be ground using a ring pulverizer to decrease the particle size below 75 μm. Care should be exercised to choose a ring pulverizer that will not contaminate the sample with the analytes of interest. The analysis of metals should be done using inductively coupled plasma-mass spectrometry (ICP-MS), inductively coupled plasma-emission spectrometry (ICP-OES) or inductively coupled plasma-atomic emission spectrometry (ICP-AES). Generally, ICP-MS is used to determine concentration levels in parts per billion and below while the ICP-AES is used to determine levels in parts per million and higher.

Field sampling reproducibility of composite sampling should be subject to QA and QC requirements such as those traditionally required to demonstrate laboratory analytical reproducibility. Field replicates provide a measure of the total error or variability of the data set. The sampling plan must provide for enough replicate QC sampling to obtain the required precision. As a rule, it is recommended that triplicate composite sampling be collected for at least 10% of all the DUs. Whenever possible, the triplicates should be collected by three teams to validate the absence of bias. All this will support collecting representative data and to make informed tailored decisions.

3.4.4 How Clean is Clean?

Some countries have adopted either thresholds or guidance soil concentration values for MCs in the soil surface. These thresholds represent guidance for range managers on how much MC can be tolerated in the surface soils of RTAs. These soil concentrations will trigger decisions such as to pursue live-firing in each area or otherwise, or to undertake remedial or mitigation actions. If site-specific soil concentrations exceed the guidelines, quantitative risk assessment and/or risk management measures can be applied.

The US EPA uses an ecological soil screening level approach, and some reference values are now available (https://www.epa.gov/risk/ecological-soil-screening-level-eco-ssl-guidance-and-documents) for main MCs. In Canada, DRDC tasked the Canadian National Research Council (CNRC) to develop explosives soil concentrations for military training sustainability. The objective was to establish environmental or ecological and human health-based guidelines to support environmentally sustainable RTAs. It was done by conducting ecological testing using various soil receptors and by evaluating and compiling data from the literature. Soldiers in various training conditions were considered as the human receptors. The generic environmental condition of RTAs was also taken into consideration. The guidelines were first published in 2006 and then updated in 2011 [26]. Environmental soil concentrations for ensuring military training sustainability are the following: TNT (9.6 mg/kg); RDX (7.7 mg/kg); HMX (89 mg/kg); 2,4-DNT (6.7 mg/kg); 2,6-DNT (10.6 mg/kg), and NG (54 mg/kg).

Finland also recently published guidance values for explosive residues, where soil screening values are risk-based and regulated by a governmental decree. The values are determined through a risk assessment framework for land-use options. A risk-based framework was used to determine generic health-based maximum acceptable concentrations (MACh) for the most common explosive substances (TNT, RDX, HMX) as well as some of their degradation products and impurities (2,4-DNT, 2,6-DNT, 1-3-5 Trinitrobenzene (TNB), and 2-and 4-amino-DNT). The MACh guidance values for TNT, RDX, and HMX are, respectively, 496, 182, and 30 000 mg/kg [27]. These values can be used for range management, keeping in mind that they are not valid when other exposure routes are considered, such as drinking and eating site-specifically contaminated water and food. Also, if the site is in a classified groundwater area, quantitative risk assessment including fate and transport modelling should be applied.

3.4.5 Risk to the Receptors Through the Transport of Munitions Constituents

Once deposited on soils in RTAs, MCs may dissolve in water, experience complex interactions with soil constituents, and/or migrate through subsurface soil leading to groundwater contamination. They can also migrate to surface water bodies through rain and snow melt. When in contact with soil, these chemicals are subject to several abiotic (hydrolysis, photolysis, reactions with metals or soil organic matter) and biotic reactions (biotransformation, mineralization), both in the solid and in the aqueous state. The main physicochemical properties to consider predicting their fate and transport are the aqueous solubility (S_w), the octanol/water partition coefficients (K_{ow}), and the soil sorption constants (K_d), as they provide insight into the fate of chemicals in the environment and the risk associated with their open-field use and applications. These parameters influence their interactions with soil and their chemical and microbial transformation routes in the environment. Knowing the reaction routes help understand their fate, ecological impact, and how to enhance *in situ* remediation. It is critical to study the fate and transport of the pure ingredients, as well as the explosives' mix formulations, as the combination of ingredients often plays a major role in governing the fate of the ingredients. This is demonstrated very well in propellant residues, in which NG will remain for years in RTA surface soils; whereas if dispersed as a pure compound, it would have degraded very rapidly (within a few days) to benign end products.

Another aspect of the fate of MCs is the physical state in which particles were deposited. The mass of the scattered materials from live-firing depends on the type of rounds and the way they are detonated: high-order, low-order (partial), or blow-in-place (BIP) detonations. At the impact area, high explosives are scattered on to soils either from high-order detonations in very fine dust down to nanometres in size, to centimetre-size chunks in low-order detonations. If the round does not detonate and becomes a UXO, all the explosive charge may eventually be released into the environment if it is not removed or destroyed.

At the FPs, propellant residues are generally deposited as fibrous material of partially burnt and unburnt particles which are deposited on top of the soil surface. The shape of the original propellant grain and the presence or absence of a perforation or hole dictates the fate of the residue. In general, propellants are based on an NC matrix, which limits the availability of the embedded components, such as 2,4-DNT and NG. Propellant residues can be deposited in front of the gun mouth in artillery live-fire training due to incomplete combustion in the gun, or behind the firing position in shoulder-fired rocket launchers. Another source of propellant in the surface soil is the burning of excess artillery propellants near firing positions. The fate of propellant ingredients varies with the degree of decomposition of the NC-based propellant matrix.

The fate and transport of MCs are now much better understood as huge research efforts have been dedicated to this in recent years [18, 28, 29]. The following general properties can be drawn from the numerous studies conducted over the years:

- RDX is, in general, not retarded, is highly mobile, and has a somewhat high water solubility (42 mg/l) which drives a higher risk to groundwater and surface water.
- TNT is, in general, retarded and may transform into more than 20 metabolites that are also retarded in the soil profile, presenting a lower risk to the water bodies and ecological receptors.
- HMX is, in general, not retarded, has lower water solubility than RDX (4 mg/l), and has proved to have a high residence time in surface soils; so HMX represents a lower risk to groundwater while it is higher risk to surface soil ecosystems.
- NG, NQ, and 2,4-DNT are, in general, embedded into an NC matrix, and thus of limited bioavailability and high residence time in surface soils. One exception is the post-burn excess propellants where NC burns preferentially leading to mobile 2,4-DNT.
- Perchlorate is very stable in the environment, highly mobile, and presents a high risk to groundwater and a lower risk to surface soils and surface ecosystems.

Human receptors can be at risk if they are exposed to the surface soils through wind and dust, such as military users at FPs. They can also be exposed through drinking of groundwater or through crop irrigation. A lot of training ranges are surrounded by either crop-growing areas or housing developments, which have been built very near military installations with the growth of the population. Therefore, the reaction time to avoid having MCs migrating out of the range and presenting a measurable risk to the surrounding population has decreased, stressing the need for a better surveillance of the surface water and groundwater bodies in RTAs. The potential for MCs to travel to water bodies in RTAs varies from quite low probability for NC-embedded propellant residues to a very high probability for RDX and AP, which are both soluble and mobile. Therefore, it is imperative to monitor both surface and subsurface water body quality.

In Canada, great effort was dedicated to the hydrogeological characterization of RTAs. Hydrogeology provides detailed information on the quality and flow direction of surface water and groundwater, on the water table depth, and on the various types of soil on which the ranges are built. In RTAs, a stepwise approach was taken in sequential phases over a period of three to four years. The number of wells installed per phase may vary between 15 and 20 and is combined with a detailed surficial and three-dimensional geological survey. This stepwise approach allows the best localization of wells and optimizes the process. One major concern for the installation of wells in UXO-contaminated ranges is the possibility of encountering buried UXO during well installations. Thus, before drilling, it is imperative to conduct UXO avoidance activities to ensure the safety of the drilling crew. This is done by qualified personnel who clear pathways to proposed sampling locations, usually using magnetometers. The pathways must be wide enough for safe passage of the drilling equipment; and, in general, a sufficiently large area will be cleared at the sampling location to allow the drilling equipment to manoeuvre properly. At all drilling locations, downhole avoidance techniques are required, and 0.5-m interval needs to be cleared using a magnetometer prior to further advancement of the drilling equipment. Another concern is that the wells will be destroyed as gunners use them for target practice. The installation of flush-mounted wells has eliminated this issue, and these types of wells have been installed within impact ranges across Canada with success.

There are multiple benefits to performing detailed hydrogeological characterization of RTAs. First, it gives timely information on the groundwater quality underlying the RTAs and allows range managers to learn if there is a problem with ground water quality, and so to act before the problem becomes out of control. It gives a picture of RTA's vulnerability through the establishment of geological maps, and informed decisions can be taken on whether to hold an activity in specific locations. As an example, antitank ranges are known to build up concentrations of concern of HMX, while demolition ranges lead to the accumulation of RDX, and, therefore, they shall not be located over highly vulnerable aquifers. The hydrogeological data collected allows the preparation of several thematic maps such as hydraulic head and surficial geology maps, groundwater flow maps which in turn allow contaminant transport modelling. A range conceptual model can be built, following the geological model and from the knowledge of the environmental fate of MCs. The risk analysis is completed by building hazards, aquifer vulnerability, and risk maps. The vulnerability map reflects the vulnerability of a given aquifer to surface contamination. The hazard for a range is evaluated with an index system and based on residue deposition intensity, on the environmental risk of each type of MC used, and on the impacted surface area. The frequency index, the environmental risk index, and the surface area of deposition index are multiplied together to give the hazard value for each of the MCs used for a given range. The aquifer vulnerability and the hazard maps are combined to generate the overall risk maps. All these thematic maps are powerful tools for range managers to promote range sustainability. Finally, performing a detailed hydrogeological RTA characterization gives high evidence of due diligence, and will be very well perceived by the community living in proximity to RTAs, limiting adverse encroachment. In conclusion, it is strongly recommended to perform hydrogeological characterization of RTAs as many short-, medium-, and long-term benefits clearly outweigh the mobilization costs.

3.5 Methodology for the Precise Measurements of MC Sources

3.5.1 Explosive Footprints in Impact Areas

Most of the heavily used impact and demolition areas show a build-up of explosive concentrations in the surface soils. To predict this and assess the related risk, it was vital to develop a methodology to measure the levels of explosive remaining post-detonation. Considering the vast amount of energy and disturbance of detonation processes, the development of a methodology to measure how much explosive remains post-detonation was not straightforward. CRREL and DRDC Valcartier succeeded in developing such a method through SERDP programmes that enables the calculation of a reproducible estimate of energetic residues mass deposition from many commonly used weapon systems [21]. The ideal set-up to measure the footprint of detonations is to conduct them in conditions as near as possible to the actual scenario of use. Therefore, our efforts were directed towards measuring the footprint in the open environment. It took years to deliver such a reliable and precise method, as detonation processes generate high temperature, pressure and disturbance, and, in general, lead to forensic levels of residues at detonation points. Various media were studied at detonation points, including soil surfaces, open range detonations, clean sand surfaces, and snow. Only one medium allowed the development of a reliable method, using pristine snow covers. Our approach involves conducting detonations at discrete detonation points, over a bed of pristine snow, and collecting the residues in a systematic approach over the deposition area, visible on the snow surface. The detonation plumes are delineated, their areas are measured, and sampling is conducted as soon as possible, using a specific sampling pattern. Figure 3.2 shows two sampling teams in the areas that were delimited on the basis of post-detonation soot pattern. The smaller circle is defined as the deposition area, while the larger circle is an area that is sampled to ensure that the deposition was large enough to collect most of the residues.

By melting the snow, the residues are dispersed in an aqueous medium, easy to concentrate and analyse (Figure 3.3). The aqueous fraction is filtered and both

Figure 3.2 Sampling post-detonation residues over snow.

Figure 3.3 Snow sample filtration.

soluble and insoluble fraction area analysed, leading to an insight into the future fate of the constituents. The use of snow has many advantages; however, it limits the windows in which tests can be conducted and it is strongly weather-dependent. The results obtained from the water extracts and from the filter solid residues are then integrated over the whole plume, which leads to the quantity of each analyte deposited per plume. The deposition rate can be calculated by dividing the amount deposited on the snow profile by the quantity of analyte in the round and multiplied by 100 to get a percentage value. Finland and Great Britain have shown interest in the North American methodology, and teams were sent to Alaska to participate in a winter deposition trial. A review paper on explosive deposition rates for various munitions has been presented at a NATO RTO AVT-177 symposium [21].

To illustrate the data now available by applying this methodology, Table 3.2 presents a data set that was obtained from the BIP detonations of conventional munitions items and was extracted from a review paper published by Walsh [30].

The method is labour-intensive and weather-dependent, although it generates relevant and critical data on explosive deposition rates and on related detonation efficiency. This further allows the evaluation of the environmental risk in target impact areas, and will soon play a role in munitions Environmental Occupational Health and Safety Assessment in Canada as well as a decision-making decision tool for munitions acquisition for which Canada has adapted the methodology for its own needs [31]. As an example, a munition showing an unacceptable deposition rate of toxic and bioavailable compound could be rejected by the acquisition process.

Returning to the source of explosives in impact areas, the following sources were studied over several years: high-order detonations, low-order detonations, UXO BIP, UXO shell cracking, and UXO corrosion [12–14]. High-order detonations normally happen when the rounds function as designed and they are defined as detonations that reach the desired pressure and detonation velocity. In general, the quantities of explosives deposited from high-order detonations are very small, almost at the forensic levels. The quantities are spread over large areas, and do not lead to the build-up of concentrations of concern of explosives with time. Low-order detonations might happen in various scenarios in live-firing events. A fraction varying from 1% to 50% of rounds might generate low-order detonations or UXOs. The failure rate of munitions depends on the type of round; and, in general, artillery rounds have a malfunctioning rate around 1–5%, while that of antitank rockets can be as high as 50%. A low-order detonation is defined as a detonation that does not reach the maximum detonation pressure and temperature. While high-order detonations deposit nanograms to micrograms of fine explosive dust, low-order detonations deposit gram quantities of explosives from fine dust to large chunks, representing an important source of MC. A third source of explosives in RTAs is in the BIP of UXO, a regular activity conducted by EOD experts to minimize the safety risks in RTAs. EOD teams regularly operate a UXO BIP operation that consists of detonating C4 blocks

Table 3.2 Deposition of RDX and HMX and related detonation rates (DR) following the blow-in place of C4 blocks and of 81-mm mortar and 105- and 155-mm artillery rounds.

Item	RDX deposited (mg)	HMX deposited (mg)	Global DR (%)
81 mm	130	23	1×10^{-2}
105 mm	41	9	2×10^{-3}
155 mm	8	4	2×10^{-4}
C4 block	12	7	3×10^{-3}

near the UXO without moving it. The impact of disposal operations conducted on RTAs by detonating UXOs with C4 is now better understood. Both the C4 donor charge and the UXO generate the deposition of MC at the detonation point. Therefore, locations where intense BIP has been conducted have a high probability of presenting measurable concentrations of MCs both from C4 and from the UXOs.

Finally, another source of explosives in RTAs was identified through the quest for understanding of the explosive contamination pattern. Surface UXOs in impact areas are susceptible to being hit by razor-sharp flying fragments from proximity high-order detonations. Designed experiments demonstrated that this phenomenon is easy to achieve and led to a gram to kilogram quantities of explosives in the surrounding environment [32]. Surface to near-surface UXOs exposed to other rounds that explode nearby represent an important source of explosives in the surface soils. The broken shells can release as much as the totality of their explosive content in the environment. This source of explosive residues stresses the importance of regular clearance of surface UXOs in RTAs.

On a longer-term perspective, corrosion of the munitions casings also represents a source of explosives in the environment and a related risk to the underlying groundwater. The corrosion rate is a complex phenomenon, which depends on the type of projectile (composition, presence/lack of coating, etc.), on soil conditions, on heating/cooling and wet/dry cycles, on soil physicochemical characteristics, and on many other parameters. It is assumed that corrosion represents a long-term source, which is still undefined, and that most of the risk is not related to surface soils but to groundwater. A study conducted in Canada, under contract on the corrosion of empty medium- and large-calibre projectiles [33], indicated that the estimated lifetime before perforation of the coated and uncoated C139 105-mm projectile used in the Canadian Armed Forces was at least 100 and 70 years, respectively. Mortars' estimated lifespan before perforation was slightly shorter, leading to at least 60, 40, and 59 years for the C70A1, the C110, and the TGM 49A4, respectively.

The conclusions from RTA characterization and from explosive deposition studies are that a forensic quantity of explosives is deposited when high explosive rounds function as designed; BIP detonations deposit a greater percentage of residues when compared to a round that would have functioned as designed; the largest explosive residue's source is low-order detonations, as 100 000 high-order detonations deposit the same amount of explosive than one low-order detonation; surface UXO cracking, caused by UXO being hit by fragments, may expose 100% of the explosive filler to the environment and represents an important source of contamination.

3.5.2 Firing Positions

Several environmental assessment studies have shown that residues coming from the incomplete combustion of gun propellant accumulate as solid fibrous particulates in front of FPs of guns, from small arms to large calibres [8, 10–14]. The nature of residues that are deposited at the gun mouth always involves an NC matrix, with either NG or 2,4-DNT embedded. Accumulation of NG and

2,4-DNT at firing points is cumulative over the years, as the NC matrix protects them from degradation and dissolution processes. A comprehensive review paper on propellant deposition rates for various munitions was presented at a NATO RTO symposium [22].

A huge body of data on propellant deposition rate was accumulated over the years in North America. To measure the propellant deposition rate, surface soils have been used, through multi-incremental soil sampling pre- and post-firing. This was soon proved unreliable, due to the deposition of trace levels of residues, which were very hard to extract from complex environmental matrices, and because of the interference from past contamination of the surface soils. Two other methods have been successfully used to measure the accumulation of propellant residues. In Canada, particle traps were used to collect the propellant particles and they were placed at the gun mouth, in a fan shape or any other relevant deposition area, such as illustrated in Figure 3.4. After trap placement, a given number of rounds are fired to produce the accumulation of measurable traces of propellant residues. This method has been successfully used to measure the deposition rate of propellant residues from a large variety of calibres. One advantage of using traps is the possible matching of the deposition pattern with the distance from the gun mouth. This is illustrated by a trial that was conducted with small arms in Figures 3.4 and 3.5. Figure 3.4 illustrates the trap collection system used for this trial, while Figure 3.5 presents the mass of residues deposited as a function of the distance from the gun mouth. This showed that most of the deposition area for this specific calibre was around 2 m from the gun mouth. This means that with time, the soil concentrations of propellant residues at this distance will grow, potentially to concentration of concerns in relation to human health. In fact, when soldiers practise live-firing with these calibres, they sometimes lie down on the soil surface, and their face is directly over the most contaminated area. It was by using traps that this health risk could be highlighted.

Another method that can be used in winter is live-firing over a bed of pristine snow, conducting the same type of sampling as described in Section 3.5.1. Most of the deposition trials conducted in the United States were conducted

Figure 3.4 Sampling small arms propellant residues using traps.

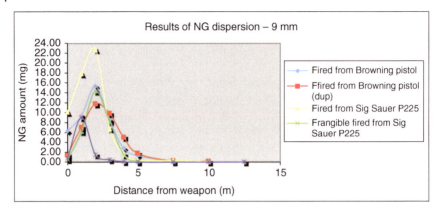

Figure 3.5 NG concentrations with the distance from a 9-mm pistol.

over pristine snow. It has the advantage of sampling the whole deposition area, using a more systematic matter. To compare the trap and snow methods, a trial was done in Canada, where both the trap and the snow were applied to monitor the residues deposited in the tank live-firing, as illustrated in Figure 3.6. The conclusion was that both methods led to the same estimate of propellant residues deposition [34, 35].

Another major source of propellant residues identified in RTAs comes from the burning of excess artillery propellant charges. Artillery guns use a propelling charge system composed of multiple charge bags to fire projectiles giving predetermined range performance. Following a gun firing operation, discarded propelling charge increments were open burnt near the gun position on the soil surface or on snow/ice in the winter. Contrary to the particles deposited at the gun mouth, which leads to a very slow dissolution rate of 2,4-DNT or NG, open burning leads to a highly leachable fraction of 2,4-DNT which is rapidly brought to the groundwater because the burning process destroys the NC protective layer.

Overall, the following areas can be considered potentially contaminated by propellant residues in RTAs' FP: in front of gun FPs from small arms to 155-mm calibres; behind and in front of anti-tank rocket FPs, and in excess propellant field expedient burning sites.

Figure 3.6 Winter trial in Canada to compare the (a) snow and (b) trap methods.

3.6 Tailored Management Practices: Mitigation and Remediation

The knowledge gained in MC dispersion from surface soil, hydrogeological characterizations, and source term evaluation is critical in order to address the problems identified and translate into design and management practices that will mitigate and attenuate range contamination. Many types of mitigation measures can be implemented to remediate or mitigate identified problems. Physical measures such as contaminant containment are suitable whenever the detonation processes involved would not destroy the containment systems. Chemical or biological treatment is another option when levels of contaminants have reached accepted guidelines or thresholds, or when the contamination lies over a sensitive area. Upstream, ranges can be designed in a sustainable manner to minimize the environmental footprint of live-fire training. The implementation of mitigation methods or permanent solutions to reduce and/or prevent environmental contamination in ranges is now being studied in many NATO nations. A NATO task group, AVT-291, focuses on innovative technology and design practices that will help ensure that the accumulation of munitions constituents does not lead to irreversible adverse effects on the environment. The measures and concepts discussed by the group are anticipated to result in a set of technical recommendations that will enable the minimization of live-fire training environmental impacts on operational training ranges. Section 3.6.1 presents two examples of actual mitigation measures being implemented, with the objective of minimizing MC footprint.

3.6.1 Mitigation Measures

One of the best options for sustainable live-fire training is the development of green munitions which remove the initial source of contaminants. This option is extensively covered in Chapter 4 of this book. The two mitigation measures presented here operate at each end of the spectrum. In Section 3.6.1.1, the approach is to monitor the problem at the source in real time, to avoid any further dispersion of environmental contaminants, thus decreasing the risk of migration towards sensitive receptors. In contrast, in Section 3.6.1.2, the approach is to remediate the surface soil from a contaminant that has a very low probability of reaching receptors through percolation, while its accumulation in surface soils could eventually lead to a risk to human health for military users.

3.6.1.1 Analytical Tool and Adsorption Method for MCs in Aqueous Samples

RTAs are contaminated with explosive residues and some are transported from the surface to the groundwater. The current procedure for measuring energetic materials in water samples is tedious and involves sampling, expedition, extraction, and laboratory analysis. An R&D project was initiated aimed at developing a cost-effective, rapid, and *in situ* field monitoring technology that will identify and measure energetic materials in water samples. The system is based on surface plasmon resonance and is led by the Montréal University, in collaboration with the University of Quebec. A joint application was made to the Natural

Sciences and Engineering Research Council of Canada. The work allowed the construction of a prototype, the field testing of the technology, and the determination of the performance of the technique for *in situ* field screening and monitoring of energetic materials. The technology was tested in the field using various groundwater well-water samples and led to the detection of RDX within 30 minutes, with high precision and accuracy. The correlation with the US EPA 8330b method was greater than 95% [36]. This technology will be further developed in the coming years, as it will represent a powerful real-time tool for better range environmental management.

Another ongoing project is the development of passive samplers for MCs in water bodies which will optimize the surveillance of the water bodies' quality in RTAs [37]. Passive diffusion samplers may provide a highly valuable tool to assess the presence of energetic constituents in all types of water bodies. They could be placed in surface water bodies or groundwater wells for a few weeks, accumulate MCs on their surface, and extracted to give the total amount of MCs that transited through the well over the sampling time. This tool will allow the integration of seasonal fluxes in the measurement of energetic contaminants, which is very difficult to assess otherwise. In Canada, the highest flux of contaminants is released in the spring, upon snow melt. Using the passive samplers in this season would be highly relevant. This will in turn help in developing more refined and efficient training range water characterization, supporting much better human health risk and environmental risk assessments. The combination of *in situ* analysis and passive samplers could also be very helpful to follow remedial actions in water bodies.

3.6.1.2 Thermal Treatment of Shoulder Rocket Propellant-Contaminated Surface and Subsurface Soils

An important source of contamination identified in RTA is the accumulation of propellant residues behind the firing position of antitank ranges, due to the back blast of the live-firing [38, 39]. This represents a threat to the military units in the contaminated area and might be exposed to NG vapours or propellant particles through dust/inhalation. A search for an affordable and easily applicable *in situ* technology was made.

The selected technology was the thermal treatment of surface soil propellant residues using indirect infrared heat emitted by a hot steel plate (Figure 3.7).

Figure 3.7 Thermal treatment of surface soils.

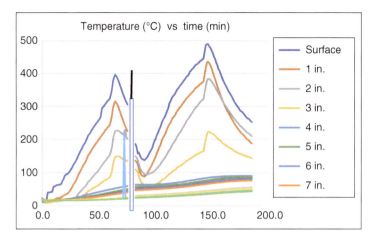

Figure 3.8 Surface and subsurface soil temperatures with time.

Very good results were obtained when using this type of thermal treatment. The surface soil temperatures were brought up to 460 °C and the heating successfully penetrated the soil profile down to 15 cm, reaching a temperature high enough to decompose the nitrocellulose/nitroglycerine matrix. Figure 3.8 shows temperatures measured with probes buried in the soil profile, with a cooling phase after 60 minutes followed by another heating cycle. Soils were effectively remediated, from concentrations ranging from thousands of milligrams/kilograms to non-detected after treatment [40].

3.7 Emerging Constituents

By conducting a thorough assessment of the environmental health of military live-fire training, range managers now have the knowledge and tools to understand and minimize MCs from the conventional munitions inventory. To ensure that there is no repetition of mistakes from the past and to be in a state of readiness for active range sustainability, we must foresee which are the emerging MCs and study their environmental properties, before they are fielded. Good examples are new and emerging insensitive formulations. NATO countries are transitioning from conventional high explosives such as Composition B to insensitive high explosive (IHE) munition systems containing insensitive formulations to reduce the risk of accidental detonations and protect military personnel.

Canada is also heading in this direction and IHE systems have been either considered for acquisition or acquired. In North America, a new family of IHE-compliant melt-cast formulations has been developed by the Armaments Research, Development and Engineering Center (ARDEC), Picatinny Arsenal, New Jersey. These formulations are based on two less-sensitive explosives: dinitroanisole (DNAN) and nitrotriazalone (NTO). These IHE formulations react better to unplanned stimuli, although they could lead to lower detonation efficiency or higher MCs' deposition rate in the environment [41, 42]. To determine the detonation efficiency of the emerging IHE rounds, trials were conducted over a four-year period (2014–2017) to measure the deposition rate of DNAN, NTO, and HMX following both high-order and BIP detonations of actual rounds. Four insensitive high-explosive formulations have been tested: PAX-21, PAX-48, IMX-101, and IMX-104 [43–46]. The PAX-21 trials indicated a very high deposition of ammonium perchlorate, which would have led to the contamination of aquifers and potentially to high liability and costly remedial actions. White papers were issued both in Canada and the United States, and the use of these munitions is now restricted. The detonation of IMX-101 and IMX-104 rounds resulted in high deposition of NTO and NQ in BIP scenarios, which are both highly water-soluble compounds. This highlights the importance of measuring MC deposition rates prior to deployment. The environmental, ecological, and human health properties of emerging reduced vulnerability constituents and formulations are also being studied in Canada and the United States (and elsewhere), to better understand their associated environmental and human health risks. The toxicity of NTO is an issue, especially considering the amount that remains following a detonation [47].

Insensitive munitions were constructed to resist external stimuli such as bullet impacts or fire, and, because of that, they can be harder to detonate. However, there is a price to pay since this insensitivity is translated in a less efficient detonation, lower performance, and increased difficulty to dispose of UXO by a BIP procedure. The more insensitive the munitions are, the less efficient they can become and the more they deposit residues. In the case where the IHE constituents are toxic, live-firing of insensitive rounds in our RTAs will represent an environmental risk. This raises the question of trading munitions' insensitivity with efficient detonation processes and illustrates the need to work on the optimization of detonation and ignition.

Other emerging MCs will also be fielded in the future, as new, better, and insensitive formulations are developed. MCs that might come to the field are, for example, nano-explosives or nano-metals, which have recently emerged in the literature. Their nanometre size could lead to very high solubility or rapid transport through colloidal particles. Other new and emerging formulations will soon be produced by three-dimensional printing, with potential groundbreaking properties. It is critical to assess the potential environmental footprint and fate of these emerging formulations by conducting systematic tests prior to fielding them in RTAs.

3.8 Conclusion

This chapter promotes the importance of a deep understanding of MC dispersion, fate, and impacts in military live-firing ranges. There is a paradox in measuring the environmental footprint of items that were designed to kill, and very often the reactions induced by this paradox are diametrically opposite to this objective. In fact, in the past 50 years, most of the NATO stockpile was fired in our own countries to ensure the readiness of our troops. Therefore, if there is an environmental risk associated with this activity, it resides in the heart of our priceless training assets and it could trigger their closure. Losing these assets would be a catastrophe for most NATO nations, as opening new ranges is almost impossible when considering the encroachment with civilian developments with the growth of the population, not to mention the 'not in my backyard' syndrome.

Twenty years ago, there was no awareness of the risk associated with the accumulation of MCs in ranges nor methods and tools for representative assessment. The journey to range sustainability was not straightforward, and a lot of resistance had to be overcome to convince the military users of the benefits of land assessment, promoting risk management. Sampling for MC from multiple sources of munitions in a variety of range types represented a great challenge. R&D demonstrated that their heterogeneous pattern drove the need for specific approaches, involving the collection of a high mass and number of subsamples, combined with proper soil processing. There was also a lot of resistance from environmental communities to the application of this approach, as it is not used for other types of contaminants and it brought a perception of dilution. The commercial laboratory also showed resistance to the recommended sample treatment, as it was considered costly and labour-intense. Again, it is critical to educate these communities to understand the importance of applying the designed protocols correctly to obtain valid data. The cost of not doing so, and of getting wrong results, is much higher. Specific protocols were also designed to measure the precise outcome of detonations and propulsion processes. Small- and large-scale fate and transport of conventional MCs were studied, and, in Canada, the numerous benefits of hydrogeological characterization of RTAs were clearly demonstrated. The priceless information obtained from all these efforts allows the design of an action plan to mitigate or eliminate the identified problems. Many countries have already begun to implement solutions to achieve sustainable training and it will grow. The process applied to MCs of our conventional stockpile must be applied to new emerging constituents and formulations to avoid repeating mistakes from the past and ensuring the viability of our ranges. Leadership and vision towards sustainable training have risen in importance in most NATO nations and shall allow continuous training of our allied forces.

References

1 Thiboutot, S., Lavigne, J., Ampleman, G. et al. (1994). Energetic compounds: application to site restoration. In: *Proceedings of the AGARD 84th Symposium, Propulsion and Energetic Panel on Environmental Aspects of Propulsion Systems*, AGARD-CP-559, 40-1–40-6. Norway: NATO Science and Technology Organization.
2 Ampleman, G., Thiboutot, S., Lavigne, J. et al. (1995). Synthesis of 14C-labelled hexahydro-1,3,5-trinitro-1,3,5-triazine (RDX), 2,4,6-trinitrotoluene (TNT), nitrocellulose (NC) and glycidyl azide polymer (GAP) for use in assessing the biodegradation potential of these energetic compounds. *J. Labelled Compd. Radiopharm.* 36 (6): 559–563.

3 Thiboutot, S., Ampleman, G., Greer, C. et al. (1996). Assessment of soils contaminated with RDX and TNT. In: *Proceedings of the 27th International ICT Conference*, vol. 114, 1–14. Karlsruhe, Germany: Energetic Materials and Manufacturing Technologies.

4 Hawari, J., Greer, C., Jones, A. et al. (1997). Soil contaminated with explosives: a search for remediation technologies. In: *Challenges in Propellants and Combustion 100 Years After Nobel* (ed. K.K. Kuo), 135–144. New York: Begell House Inc. Publishers.

5 Thiboutot, S., Ampleman, G., Dubé, P. et al. (1998). Insight for the characterization of explosives contaminated sites. In: *Proceedings of the 29th International ICT Conference*, 127-1–127-14. Karlsruhe, Germany.

6 Robidoux, P.Y., Hawari, J., Thiboutot, S. et al. (1999). Acute toxicity of 2,4,6-trinitrotoluene in earthworm. *Ecotoxicol. Environ. Saf.* 44 (3): 311–321.

7 Thiboutot, S., Ampleman, G., Dubé, P., and Hawari, J. (1997). Protocol for the Characterization of Explosives-contaminated Sites. *DREV-R-9721*.

8 Thiboutot, S., Ampleman, G., Gagnon, A. et al. (1998). Characterization of Antitank Firing Ranges at CFB Valcartier, WATC Wainwright and CFAD Dundurn. *DREV R-9809*.

9 Jenkins, T.F., Grant, C.L., Walsh, M.E. et al. (1999). Coping with spatial heterogeneity effects on sampling and analysis at an HMX contaminated antitank firing range. *Field Anal. Chem. Technol.* 3 (1): 19–28.

10 Thiboutot, S., Ampleman, G., Lewis, J. et al. (2002). Assessment of the environmental impact of the firing activity on active Canadian firing ranges. In: *Proceedings of the 33rd International ICT Conference*, vol. 55, 1–12. Karlsruhe, Germany.

11 Jenkins, T.F., Hewitt, A.D., Grant, C.L. et al. (2006). Identity and distribution of residues of energetic compounds at army live-fire training ranges. *Chemosphere* 63: 1280–1290.

12 Pennington, J.C., Jenkins, T.F., Ampleman, G. et al. (2006). Distribution and Fate of Energetics on DoD Test and Training Ranges. Final Rep. *ERDC TR-06-13*. US Army Corps of Engineers.

13 Jenkins, T.F., Ampleman, G., Thiboutot, S. et al. (2008). Characterization and Fate of Gun and Rocket Propellant Residues on Testing and Training Ranges. Final Rep. *ERDC Tech. Rep. TR-08-1*. US Army Corps of Engineers.

14 Thiboutot, S., Ampleman, G., Brochu, S. et al. (2012). Environmental characterization of military training ranges for munitions-related contaminants. *Int. J. Energetic Mat. Chem. Prop.* 11 (1): 15–57.

15 Cumming, A., Hagvall, J., Thiboutot, S. et al. (2010). *Environmental Impact of Munitions and Propellant Disposal*. NATO Unclassified, 1–86. NATO Science and Technology Organization. ISBN: 978-92-837-0105-7.

16 NATO AVT-177 Symposium, (2011). *Munition and Propellant Disposal and Its Impact on the Environment*. Edinburgh, UK: NATO Science and Technology Organization https://www.cso.nato.int/ACTIVITY_META.asp?ACT=1608.

17 NATO AVT-243 Symposium, (2014). *Next Generation Greener Energetics and Their Management*. Belgium: NATO Science and Technology Organization https://www.cso.nato.int/Detail.asp?ID=7212.

18 Walsh, M.R., Walsh, M.E, Voie, O. et al. (2015). Munitions Related Contamination: Source Characterization, Fate and Transport. *NATO AVT-197 Rep.*, NATO Unclassified, releasable to Partners for Peace, Finland and Sweden, 1–88.
19 NATO AVT-244 Symposium, (2015). *Munitions Related Contamination*. Czech Republic: https://www.cso.nato.int/Detail.asp?ID=7693.
20 NATO AVT-249. Cooperative demonstration of technology (CDT): munitions related contamination – military live-fire range characterization. https://www.cso.nato.int/activity_meta.asp?act=6142.
21 Walsh, M.R., Walsh, M.E., Taylor, S. et al. (2011). Explosives residues on military training ranges. In: *Proceedings of the NATO AVT-177 Symposium on Munition and Propellant Disposal and Its Impact on the Environment*, 27-1–27-22. U.K.: Defence Research and Development Canada.
22 Ampleman, G., Thiboutot, S., Walsh, M.R. et al. (2011). Propellant residue deposition rates on army ranges. *Proceedings of the NATO AVT-177 Symposium on Munition and Propellant Disposal and Its Impact on the Environment*, United Kingdom, 24-1/24-25.
23 Goldsmith, G.S. and Fittipaldi, J. (2010). Investigation and summary of what we have learned after 12 years at Massachusetts Military Reservation (MMR) and implications for the continued use of military ranges in the United States. In: *Army Environmental Policy Institute*. http://www.aepi.army.mil/awcfellows/docs/Environmental_Impacts_of_Military_Range_Use.pdf.
24 Hagfors, M. (2013). Destruction of old expired and spoiled munitions in Finland, environmental effects of open surface mass detonations. *Proceedings of the 1st European Conference of Defence and the Environment Symposium*.
25 US EPA 8330b Method. (2015). Nitroaromatic, nitramines and nitrate esters by high performance liquid chromatography, SW-846. https://www.epa.gov/sites/production/files/2015-07/documents/epa-8330b.pdf.
26 Berthelot, Y., Sunahara, G.I., Robidoux, P.Y. (2011). Explosives Soil Concentrations for Military Training Sustainability (SCMTS): An Update. *CNRC-NRC Rep. #53355*.
27 Koponen, K. and Vesterinen, M. (2015). Development of guidance values for explosive residues. *Proceeding of the 2nd European Conference of Defence and the Environment Symposium*.
28 Sunahara, G., Lotufo, G., Kuperman, R. et al. (2009). *Ecotoxicology of Explosives*. CRC Press/Taylor and Francis Group. ISBN: 978-0-8493-2839-8.
29 Pichtel, J. (2012). Distribution and fate of military explosives and propellants in soil: a review. *Appl. Environ. Soil Sci.* 2012: 1–33.
30 Walsh, M. (2007). Explosives Residues Detonation of Common Military Munitions: 2002–2006. *CERDC/CRREL Rep. TR-07-2*.
31 Thiboutot, S. (2017). Scientific Methodology for the Precise Determination of Post-detonation Explosive's Footprint. *DRDC-RDDC-2017-R112*. Defence Research and Development Canada.
32 Lewis, J., Martel, R., Trépanier, L. et al. (2009). Quantifying the transport of energetic materials in unsaturated sediments from cracked unexploded ordnance. *J. Environ. Qual.* 38: 1–8.

33 Chavez, J. (2014). Etude du mécanisme de corrosion des munitions. Rapport B-0007995-1-MC-0001-00, Contract DVRD2013, Dessau. US Army Corps of Engineers.

34 Ampleman, G., Thiboutot, S., Marois, A. et al. (2008). Evaluation of the Propellant Residues Emitted During 105 mm Leopard Tank Live Firing. *RDDC Valcartier TR-2007-515*.

35 Walsh, M.R., Marianne, E., Walsh, M.E. et al. (2012). Munitions propellants residue deposition rates on military training ranges. *Propellants Explos. Pyrotech.* 37: 393–406.

36 Brulé, T., Granger, G., Bukar, N. et al. (2017). A field-deployed surface plasmon resonance (SPR) sensor for RDX quantification in environmental waters. *Analyst* 142 (12): 2161–2168.

37 St. George, T., Vlahos, P., Harner, P., and Helm, B.A. (2016). Rapidly equilibrating, thin film, passive water sampler for organic contaminants; characterization and field testing. *Environ. Pollut.* 159 (2): 481–486.

38 Thiboutot, S., Ampleman, G., Marois, A. et al. (2007). Deposition of Gun Propellant Residues from 84-mm Carl Gustav Rocket Firing. *RDDC Valcartier TR-2007-408*.

39 Walsh, M.R., Walsh, M.E, Thiboutot, S. et al. (2010). Propellant Residues Deposition from Firing of AT4 Rockets. *ERDC/CRREL TR-09-13*, 1–41.

40 Downe, S., Ampleman, G., Thiboutot, S., and Keys-Connell, K. (2015). Remediation of NC/NG contaminated surface soil at firing positions by open burning. *Global Demilitarization Symposium, NDIA Conference*, Parsippany, NJ.

41 Thiboutot, S. and Walsh, M. (2018). SERDP Project ER-2219. https://www.serdp-estcp.org/Program-Areas/Environmental-Restoration/Contaminants-on-Ranges/Characterizing-Fate-and-Transport/ER-2219/ER-2219 (accessed November 2018).

42 Brousseau, P., Ampleman, G., and Thiboutot, S. (2013). Residues from a detonation: are green and IM compatible? *Proceedings of the NDIA Insensitive Munitions and Energetic Materials Symposium*, USA.

43 Thiboutot, S., Ampleman, G., Diaz, E., and Brousseau, P. (2017). Detonation Efficiencies of Insensitive 120 mm Tank Round: Risks Related to Insensitive Munitions. *DRDC-RDDC-2017-R014*.

44 Walsh, M.R., Walsh, M.A., Taylor, S. et al. (2013). Characterization of PAX-21 insensitive munition detonation residues. *Propellants Explos. Pyrotech.* 38 (3): 399–409.

45 Walsh, M.R., Walsh, M.E., Ramsey, C.A. et al. (2014). Energetic residues from the detonation of IMX-104 insensitive munitions. *Propellants Explos. Pyrotech.* 39 (2): 243–250.

46 Walsh, M.R., Walsh, M.E., Ramsey, C.M. et al. (2013). Perchlorate contamination from the detonation of insensitive high-explosive rounds. *J. Hazard. Mater.* 262: 228–233.

47 Johnson, M., Eck, W., and Lent, M. (2016). Toxicity of insensitive munition (IMX) formulations and components. *Propellants Explos. Pyrotech.* 42 (1): 9–16.

4

Greener Munitions

Sylvie Brochu and Sonia Thiboutot

Defence Research and Development Canada – Valcartier Research Center, 2459 de la Bravoure Road, Quebec, QC, G3J 1X7, Canada

4.1 Background and Context

Training with munitions is critical to maintain operational readiness of the Armed Forces around the world. However, the use of munitions was shown to lead to the dispersion of energetic material (EM) residues in military ranges and training areas (RTAs) [1–3], some of which are constituents of concern because of their potential risk to sensitive receptors. While remediation and mitigation have been found effective to prevent unnecessary dispersion of EMs outside boundaries of small size ranges, such as firing positions (Chapter 3 from Thiboutot in this book) and small arms ranges [4–6], controlling the contamination directly at its source before its dispersion using greener munitions is seen as a more promising avenue to avoid extensive spreading of energetic residues in RTAs. Greener munitions would indeed reduce the need to resort to expensive and lengthy RTA cleaning or remediating as well as the necessity to develop costly site-specific range attenuation designs.

The dispersion of EM residues and unexploded ordnance (UXO) in battlefields and in overseas military bases poses additional challenges. Since there was no *a priori* environmental awareness at the time of the installation of several overseas military bases, most international agreements do not take environmental considerations into account. One of the biggest issues at the closure of such facilities is therefore to determine who is responsible to pay for the remediation. The choice of appropriate cleaning standards, a driving cost factor for contaminated sites remediation, also represents a huge challenge because each country has its own standards. Consequently, overseas base closure clean-up issues have the potential to jeopardize the relationship of involved countries. The use of greener munitions could prevent or alleviate such difficulties, in addition to attenuating environmental impacts and health hazards for civilian populations and military users.

Substantial efforts have been undertaken in the past two decades to develop greener energetics that would have less adverse environmental impacts in military RTAs and in active use or would present less health hazards for the users [7]. However, very few of these greener energetic materials have made it into an actual munition system. The incorporation of a new molecule into a munition is a very complex process. There are several aspects to take into consideration when going from the molecule, to the formulation, and then to the munition. Just for concept demonstration purpose, one has to consider, among others, the performance assessment (subcalibre, full-calibre, fragmentation test, etc.), the hazard assessment (slow and fact cook-off, bullet impact, sympathetic detonation, etc.), the integration assessment of all munitions' components (chemical compatibility, barrel wear, loadability, etc.) as well as the stability (chemical, thermal, mechanical, etc.). Potential commercialization also involves numerous crucial, costly, and time-consuming steps, such as production optimization, the safety and suitability for service assessment, as well as the munitions' qualifications.

For example, several research and development (R&D) efforts were focused on the development of lead-free bullets or primers for small-calibre ammunition. Small arms training does indeed occupy a huge portion of the military activities, because all service personnel must be qualified in the handling of a personal weapon. Millions of small-calibre rounds are fired annually, either in operational activities or to maintain the Armed Forces in a high state of preparedness. Small arms training ranges are being used extensively, which contributes to the escalation of residue accumulation on-site.

Significant lessons were learnt from these R&D efforts. For example, the ban of lead at the US Massachusetts Military Reservation (MMR) in 1997 followed its detection in the groundwater, the sole source aquifer underneath the MMR [8]. As a result, the US Army's Green Ammunition Program selected tungsten as a replacement for lead, because it was believed to be insoluble in water and non-toxic [9]. However, tungsten was later found to readily dissolve in water under specific field conditions [10, 11], and to present some environmental toxicity, even worse than lead. Therefore, the use of tungsten in MMR was suspended in 2006 [12]. The lesson learnt here is that environmental impacts should be assessed sooner in the munition development process, contrary to the current approach consisting of evaluating the environmental impacts and health hazards only at the end of the development process.

Another significant lesson learnt was documented by the Norwegian Armed Forces, which introduced unleaded Nammo ammunition into their standard weapon, the HK416. Military users soon began to complain about acute health effects such as fever, chills, nausea, headache, fatigue, muscle and joint pain, cough, and shortness of breath. A thorough investigation demonstrated that these metal fever–related symptoms were due to high emissions of copper and zinc coming from the jacket [13]. However, this particular Nammo ammunition had been intensively used before in other weapons systems without presenting any significant issues. This illustrates the fact that environmental impacts and health hazards are dependent on weapons systems, and cannot be predicted solely by the properties of the energetic or inert materials.

The development of the IM 60-mm mortar based on PAX-21 in the United States was an additional example of munition development ending in a dead end because the environmental impacts were not assessed soon enough in the munition development process. This round responded very well to all insensitive munitions (IMs) and operational requirements, and its production was undertaken on a large scale. However, trials to measure the post-detonation residues of these rounds indicated a high deposition rate of unburnt perchlorate residues at the detonation point, even in high-order detonation scenarios, leading to a very high risk for the aquifers [14, 15]. Based on these results, the production of the round was halted.

On the other hand, the replacement of perchlorate and heavy metals in some pyrotechnic applications proved successful in several munition systems. The search for a perchlorate replacement began in the early 2000s, after it was established in 1997 that the use of the M116 artillery simulator at Camp Edwards was one of the causes of environmental contamination [8], together with lead and hexahydro-1,3,5-trinitro-1,3,5-triazine (RDX) issued from other calibres. Since then, the Strategic Environmental Research and Development Program (SERDP) has funded numerous projects, mostly at the US ARDEC Pyrotechnic Research and Technology Branch and at NSWC Crane, aimed at developing appropriate perchlorate replacements displaying similar or improved performance and sensitivity compared to in-service formulations, while exhibiting less environmental impacts. Projects have focused on developing compositions for visible signal flares, smoke signals, delay compositions, and incendiary compositions. Several perchlorate-free compositions have been successfully transitioned to in-service items, such as the M115A2 ground burst projectile simulator, the M116A1 hand grenade simulator, and the M274 2.75″ rocket simulator [16]. Some of these are discussed in greater detail elsewhere in this book.

Greener munitions concepts were also explored by Canada. This chapter presents an overview of the Canadian approach, illustrated by two case examples: a Technology Demonstration Program (TDP) called RIGHTTRAC (Revolutionary Insensitive, Green and Healthier Training Technology with Reduced Adverse Contamination) and the development of an RDX-free plastic explosive as a replacement for C4.

4.2 Munitions Constituents of Concern

The main munitions constituents of concern in impact areas are RDX and perchlorates. RDX mostly originates from Composition B (Comp B) and Composition C4. Comp B, which also contains 40% 2,4,6-trinitrotoluene (TNT), is used as the main explosive filling in artillery projectiles, rockets, landmines, hand grenades, and various other munitions and has been since World War II. C4, made of 94% RDX, is used in demolition blocks. RDX is neither degradable nor adsorbed by soil and, because of its solubility in water, migrates easily to groundwater and off military property. This may trigger some serious environmental problems and become a public health concern if the groundwater is used for drinking. For example, contamination of groundwater by RDX has been identified as one of the main causes of the MMR cessation of training at the end of the 1990s. Several million dollars have been spent in groundwater remediation since then.

Perchlorates, because of their high chemical stability and their very high water solubility (from 15 to 2000 g/l, depending on the counter ion) [17], are known to be very mobile in the environment and to produce long and persistent contamination plumes when released in groundwater or surface water [17]. The MMR and Camp Edward were shut down by the US Environmental Protection Agency (EPA) in 1997 partly due to the presence of perchlorates in the surrounding groundwater, which was the main source of drinking water for the Cape Cod Area (MA) [8]. However, exposure to high concentrations of perchlorate can affect human health, particularly the thyroid gland [18] of children and pregnant women. As a result, the US EPA Interim Lifetime Drinking Water Health Advisory figure, which corresponds to a concentration of perchlorate in drinking water that is not expected to cause adverse non-carcinogenic effects for a lifetime exposure, has been established at 5 µg/l [19].

Propellants contain significant amounts of carcinogenic and toxic components, some recently forbidden in the European Union (Registration, Evaluation, Authorisation and Restriction of Chemicals [REACH], EU [20]), which could have a health impact on soldiers. Several RTA firing points were contaminated with concentrations of energetic residues above the guidelines, sometimes by many orders of magnitude. In firing positions, environmental issues are mainly due to the presence of 2,4-dinitrotoluene (2,4-DNT) and 2,6-ditrotoluene (2,6-DNT), which is an impurity of 2,4-DNT. 2,4-DNT is mainly used as a stabilizer of single nitrocellulose-based propellants in some medium- and large-calibre munitions and as a modifier for smokeless gunpowder. Due to its moderate solubility, DNT has the potential to be transported in surface water or groundwater; however, it has seldom been detected in water bodies under Canadian RTAs [21], probably because it can remain embedded in NC matrices for decades [22–24]. 2,4-DNT was, however, detected beneath open burning sites of excess artillery propellant and a field demilitarization method was developed in Canada to mitigate this risk.

4.3 Source of Munitions Constituents

It has now been established that normally functioning munitions only spread a small amount of energetic residues in the environment. However, partial (low-order) detonations constitute a significant environment concern: a single low-order detonation spreads indeed as much undetonated energetic residues as 10 000–100 000 high-order rounds [25].

The lack of detonation, leading to the production of UXO, also represents a significant issue. UXO, indeed, pose safety problems for the troops, both in training and in operations, and have to be regularly cleared from the soil surface by blow-in-place (BIP) using C4 [26, 27]. This demilitarization operation results in the generation of a larger amount of undetonated residues, issued from both the munition and the C4 itself [28–30]. UXO that are not BIP can be easily cracked open by the detonation of an incoming round [31], or, after several decades, be corroded [32], consequently leaving the entire energetic formulation exposed to weathering. Ultimately, when ranges and overseas military bases are decommissioned, UXO remain a costly problem to clean-up whenever alternative land uses are considered.

Adverse environmental impacts are also due to IMs, which are designed to react to unintentional stimuli, such as bullet or fragment impact, or fast or slow heating, in such a way as to not cause catastrophic collateral damage that impairs warfighting capability. The use and BIP of IMs, and especially the BIP, led to the dispersion of a much larger amount of undetonated energetic residues than current munitions, sometimes by several orders of magnitude [30].

The main sources of firing position contamination are partial combustion of propellant in the gun barrel and the mandatory burning of excess artillery gun propellant bags after artillery exercises. The ratio of unburnt propellant residues is dependent on weapon systems, and was found to vary from 10^{-3}% to 10^{-6}% for large-calibre munitions weapon systems to above 70% for shoulder weapon systems [33, 34]. The deposition rate resulting from the firing of mortars and small-calibre ammunition is of similar order of magnitude (generally up to 3%), although small arms have the potential for much larger residue accumulations because of their intensive usage.

4.4 Greener Munitions Development Approach

Green munitions are generally defined as munitions that do exhibit less environmental impacts without adversely impacting current performance or insensitivity level of munitions. For example, the concept of green munitions was defined by Brinck [7] as *a material designed and manufactured in accordance with the principles of green chemistry, with the minimum requirement to preserve the performance level, and safety of handling of the energetic material it is intended to replace.* Green chemistry has been defined by the US EPA as *the design of chemical products and processes that reduce or eliminate the generation of hazardous substances* (USEPA, [35]).

However, at this point in time, environmental impacts caused by the use of munitions in RTAs and in operations are unavoidable (unless we go back to throwing stones and arrows, which is quite unlikely). Yet, the risk to sensitive receptors, such as drinking wells, humans, and fauna and flora located outside the range boundaries, from toxic and carcinogenic compounds such as RDX and DNT, must be minimized in order to sustain military training. As the risk incurred by any given receptor is both a function of its degree of exposure to a particular product as well as to the toxicity of this product, the risk to receptors can consequently be reduced by using either innocuous products, or by preventing the exposure of receptors to these products. Consequently, greener munitions can be defined as munitions that do exhibit a lower degree of risk to sensitive receptors without altering either their performance or their insensitivity level. Limiting the exposure of military users and receptors outside the base boundaries in domestic and overseas RTAs, where most EM accumulation has been noticed up to now, is critical.

Given that definition, RDX cannot be considered a green constituent, because of its toxicity, its chemical stability in the environment, and its scarce water solubility which allow it to move relatively rapidly towards surface and groundwater bodies in RTAs and on battlefields. Although not a regulated substance in Canada, published guidance values for RDX in drinking water and for the protection of surface water resources are low, in the parts per billion range. Recently, RDX concentrations approaching published guidance values were detected in water bodies of interest in Canada, triggering a decision to restrict the use of RDX-based items in some sensitive areas. Thus, there is a pressing need for studying potential greener alternatives to RDX-based explosives, as the continued use of such ordnance will increase the contamination of water bodies both in and around RTAs and in operations where the controls are significantly less.

Several commercially available EMs, common and emergent, have similar or better explosive properties than RDX:

- Ammonium dinitramide (ADN);
- 1,1-diamino-2,2-dinitroethylene (FOX-7);
- Dinitroanisole (DNAN);
- Dinitrotriazalone (NTO);
- N-guanylurea-dinitramide (FOX-12);
- 2,4,6,8,10,12-hexanitro-2,4,6,8,10,12-hexaazaisowurtzitane (CL-20);
- Octahydro-1,3,5,7-tetranitro-1,3,5,7-tetrazocine (HMX);
- Pentaerythritol trinitrate (PETN);
- TNT.

With the exception of HMX and PETN, all these EMs were found to exhibit various toxicity levels when exposed to either aquatic species (freshwater microalgae and marine bacteria) or to terrestrial organisms (earthworms and ryegrass plants). For example, TNT, 2,4-DNT, DNAN, NG, FOX-7, and FOX-12 were found to inhibit algal growth at low concentrations, with TNT displaying the highest toxicity. CL-20 exhibited no toxicity for aqueous receptors because of its low solubility [36], while NTO and NQ were only toxic at higher concentrations. In addition, the nitroaromatics TNT, DNT, and DNAN proved to be cytotoxic, and DNAN genotoxic [37, 38].

Regarding terrestrial toxicity, RDX did not display toxicity to terrestrial organisms in the earthworm lethality test or in the rye grass growth. CL-20 and FOX-7 were not toxic to rye grass, but were to earthworms. 2,4-DNT, DNAN, and NTO exhibit toxicity to rye grass growth at low concentrations, while TNT and NG were toxic at higher concentrations. TNT, DNAN, and CL-20 exhibit toxicity to earthworms at low concentration, while NG, NTO, and FOX-7 were toxic to earthworms only at higher concentrations. NTO was extremely toxic to rye grass growth, probably because of the low pH of the soil NTO came in contact with (the dissolution of NTO in water led to a pH of 2.4). In addition, NG, CL-20, and RDX were also subjected to avian toxicity tests. All exhibited various toxicity levels [37].

To summarize, nitroaromatics (TNT, DNT, and DNAN) are generally considered more toxic than nitramines (RDX, HMX, and CL-20). However, DNAN is more toxic than TNT to terrestrial receptors, while TNT is significantly more toxic to aquatic species. NTO is extremely toxic to rye grass. Consequently, choosing a green EM based solely on its toxicity is not that simple, especially when other munitions characteristics such as performance and insensitivity have to be taken into account. Risks to sensitive receptors have to be considered in order to make an informed decision.

The main routes of exposure of sensitive receptors to EMs are through groundwater and surface water. Airborne particulate transportation also contributes to the dispersion of EMs, and might become increasingly important in the future in relation to encroachment and housing development approaching the limits of training ranges. However, airborne transportation was way less studied than water transportation, so this subject is not elaborated here.

Hence, the water solubility of EMs plays a critical role in the potential degree of exposure of a sensitive receptor. According to Table 4.1, which lists the water solubility of the most common EMs, PETN, CL-20, and HMX represent a very low risk to sensitive receptors, even in a case where they would exhibit a mild to moderate toxicity. On the other side of the spectrum, NTO and ADN represent a strong risk to sensitive receptors if not absorbed by soil constituents, unless they display absolutely no toxicity. All the other EMs are in between, and hence their ability to sorb to soil constituents as well as their toxicity need to be thoroughly evaluated in view of a holistic risk assessment.

With the exception of CL-20, which has a high affinity for organic soil, the above-mentioned EMs do not tend to be absorbed by soil [37] and are thus extremely mobile in the environment. However, TNT and DNAN degrade into polyamines that bind to soil content, and hence might not be available for ecological receptors. This is the reason why TNT is seldom detected in the groundwater and surface water of RTAs [21].

To summarize, based on the risk to sensitive receptors, HMX and PETN would constitute excellent choices, depending on their lack of toxicity and their low potential to reach sensitive receptors. PETN was never detected in any environmental soil or water sample out of the thousands of samples collected across

Table 4.1 Water solubility of energetic materials.

Energetic material	Water solubility (mg/l)
PETN	2
CL-20	4
HMX	5
TNT	130
2,4-DNT	200
DNAN	276
FOX-7	320
NG	1 800
NTO	16 600
ADN	3 750 000

North American RTAs, even though it is used in various military items such as datasheet and detonating cords [3, 25, 34, 39]. PETN's half-life is between 0.4 and 2.4 days, meaning that it will degrade very rapidly once dispersed in the environment [40]. The half-life value means that after this delay, half of the concentration of PETN will be transformed. In comparison, the half-life of RDX is in months, between 94 to 154 days. Also, there are no official thresholds for PETN in the German law for drinking water [41], and not in any NATO nation, to the best of our knowledge.

TNT, DNAN, and CL-20 would constitute acceptable choices, based on their low mobility in soil. Moreover, HMX and CL-20 have greater detonation velocities than RDX [42]. Unfortunately, their use is limited because of their high cost, especially for CL-20. From an environmental point of view, NTO is a potential contaminant of concern, because of its toxicity related to its low pH, coupled with its extremely high mobility. ADN is also a contaminant of concern, because of its high mobility. Moreover, its sensitivity to impact and friction is high.

The next sections present two case examples: the TDP RIGHTTRAC and the RDX-free plastic explosive C4.

4.5 RIGHTTRAC

The greener munitions concept explored by Canada, called RIGHTTRAC, aimed to show that green munitions and insensitive munitions had better properties than current munitions, and that it was feasible to implement safer weapon solutions that would ease the environmental pressure on RTAs, and decrease the health hazards for the users. The goal of the RIGHTTRAC TDP was to reach a near-zero dud rate and eliminate the potential for RDX contamination as well as the use of toxic and carcinogenic compounds. As shown in Figure 4.1, the vehicle used for this demonstration was a 105-mm army artillery munition (HE M1), currently filled with Composition B (Comp B) and using a single-base gun propellant (M1 formulation). The project tackled the three main components of a 105-mm munition:

- The fuze, to add an independent self-destruct capacity to existing fuzes in order to significantly reduce or eliminate the UXOs;

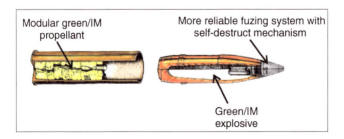

Figure 4.1 RIGHTTRAC concept.

- The gun propellant, to replace toxic or carcinogenic ingredients, to improve the IM characteristics of the propellant and to explore the concept of modular charges;
- The explosives, to replace the RDX in the main charge and to obtain an IM formulation.

IM was a safety prerequisite, and a requirement to comply with international agreements. The development work on the fuze, main explosive charge, and gun propellant ran concurrently during the first two years of the project. Work was also performed on the booster and on the primer, to ensure that both would effectively ignite the IM explosive/propellant charge [43, 44]. Two explosives and six gun propellant candidates were evaluated by performing material characterization, sub-scale IM testing, and subcalibre ballistic assessment.

The whole life-cycle cost was evaluated using a cost-efficiency analysis (CEA), which was deemed more relevant than the unit cost from a sustainable military training perspective. In addition, the aim of RIGHTTRAC was to prove the greener munitions concept by a field demonstration, excluding thereby any subsequent munitions production optimization, qualifications, and safety and suitability for service (S^3) testing, and consequently preventing the determination of a precise unit cost.

4.5.1 Energetic Formulation Selection

4.5.1.1 Main Explosive Charge

To be considered in the project, main explosive charge candidates had to be RDX-free and had to meet the current performance criteria of Comp B, which is the current explosive formulation in 105-mm munitions. One of the candidates was CX-85, a cast-cured polymer-bonded explosive (PBX) composed of a mix of hydroxyl-terminated polybutadiene (HTPB), HMX, and di(2-ethyl hexyl) adipate (DEHA), a plasticizer. The other candidate, developed and patented by DRDC, was a melt-cast formulation made of TNT, HMX, and an energetic thermoplastic elastomer (ETPE) [45–47]. Because of the existing large industrial base for the processing of melt-cast formulations, those are more common in North America than cast-cured formulations.

4.5.1.2 Performance

The conformity of GIM and CX-85 to the performance prerequisite was verified using detonation pressures and velocities, as well as plate dent tests [48]. As indicated in Table 4.2, the performance of both candidates was within 10% of Comp B. GIM and CX-85 were thus judged as qualified for further testing in RIGHTTRAC.

4.5.1.3 IM Properties

IM testing of munitions is required to assess hazardous consequences due to unintended reactions of munitions and EMs through their life cycle and is mandatory for NATO Armed Forces interoperability. IM testing must be performed on munitions according to Standard Agreements (STANAG). This

Table 4.2 Performance of GIM and CX-85 relative to Comp B.

Formulation	Density (g/cm³)	Relative VoD (% Comp B)	Relative P (% Comp B)	Plate dent (% Comp B)
CX-85	1.61	103	102	90
GIM	1.67	97	94	96
Comp B	1.68	100	100	100

P = pressure; VoD = velocity of detonation.

type of test is expensive, and moreover requires a significant quantity of energetic formulation. Consequently, IM testing on the complete rounds were planned in RIGHTTRAC only for the selected candidates. Instead, in-house small-scale IM tests were designed to compare candidates to the reference formulation and to discriminate between preselected candidates. Preliminary small- and full-scale IM tests were performed on GIM and Comp B, namely Shaped-Charge Jet (SCJ) attack, Bullet Impact (BI), Slow Cook-Off (SCO) and Sympathetic Reaction (SR) [49].

The reaction of both candidates to BI was much better than Comp B's. Indeed, neither GIM nor CX-85 detonated upon bullet impact; they rather burnt slowly, in contrast to Comp B, which detonated. All the explosive formulations passed SR, albeit with different levels of reaction: Comp B and GIM exploded (Type III), and CX-85 produced no reaction: the projectiles were simply crushed and some formulation chunks were ejected out of the shell. GIM and CX-85 performed as poorly as Comp B for the SR test. This level of reaction was not surprising, considering the severity of the threat. To the best of our knowledge, it is extremely difficult to obtain a milder reaction to the 84-mm SCJ test with the military energetic formulations commercially available. Lastly, both candidate formulations performed as well as Comp B in SCO and passed the test. Results are summarized in Table 4.3 and in Brousseau et al. [49].

4.5.1.4 Fate, Transport, and Toxicity

For formulations, a complete risk assessment would require the evaluation of the environmental and health impacts (EHI) of all the ingredients in the formulation throughout all the energetic formulation life cycle [50], from the conception of individual ingredients to their end-of-life usage, e.g. use in-service or

Table 4.3 IM tests results for the explosive candidates and reference formulation.

Formulation	BI	SCJ	SR	SCO
Comp B	I or II	I	III	V
GIM	V	I	III	V
CX-85	V	I	NR	V

NA = not available; NR = no reaction; Red = fail; green = pass.

demilitarization. A complete EHI assessment is a very complex, lengthy, and costly process [44, 50]. Known harmful compounds regulated in Canada, the United States, and the European Union (EU, GC [51, 52] and ATSDR websites) [53] are the result of a thorough research process and data processing. For compounds of interest that are not regulated, an estimation of the EHI can be performed using simple data, such as the water solubility, the soil/water distribution coefficient (K_d), and the octanol–water coefficient (K_{ow}).

Within the context of this project, where a total of eight energetic formulations were evaluated, this would have resulted in a tremendous effort and required a substantial budget. Instead, a more practical approach was adopted, which involved the evaluation of EHI only for ingredients that could leach from the formulation and thus cause a threat to the environment or the users' health.

Leaching was evaluated both under laboratory conditions using the dripping test and weathered under outdoor conditions. Results later demonstrated that the dripping test represents the worst-case scenario, and that outdoor conditions were much more representative of the actual fate of formulations in Canada. Following leaching, toxicity and transport studies were performed solely on mobile ingredients. In addition, the emission of particulate and air emissions were evaluated for propellant formulations. A recycling study was also performed to evaluate end-of-life options. Whenever possible, values published from official sources, such as the EPA, the EU, REACH, and the Canadian List of Challenge substances (EU and GC websites), were used in this project. Details are provided in the following sections and in the references therein; REACH is also the focus of Chapter 9 of this book.

GIM and CX-85 were much more resistant to dissolution than Comp B [54–56]. Dissolution tests led to the following order: Comp B > GIM > CX-85. Comp B completely dissolved in a few months on contact with dripping water, releasing RDX and TNT. Under the same conditions, neither CX-85 nor GIM fully dissolved in one year. Both candidate formulations are not even expected to dissolve completely, as suggested by the dissolution kinetics. As reported in Table 4.4, HMX was released from both formulations, TNT from GIM, and DEHA from CX-85.

Once dissolved from the formulation, the transformation and migration properties of HMX, TNT, and DEHA were as pure compounds [57, 58].

Table 4.4 Transport properties and toxicity of Comp B, GIM, and CX-85.

Formulation	Released ingredients	Terrestrial toxicity[a]	Sediment toxicity[a]	Aquatic toxicity[a]
Comp B	TNT, RDX	+++	+++	+++
GIM	TNT, HMX	+++	+++	+++
CX-85	DEHA, HMX	ND	ND	ND

ND = not detected.
a) The number of + signs is related to the number of tests for which deleterious effects were observed. A total of 10 different toxicity tests were performed.

4.5.2 Main Propellant Charge

To be considered in the project, propellant candidates had to meet the current performance criteria of the current artillery 105-mm propellant (M1). In addition, the potential candidates had to be free of nitroglycerin (NG), DNT, phthalates derivatives, and diphenylamine (DPA). Several options were considered at some point during the course of the project:

1) A single-base propellant containing nitrocellulose (NC) and an inert plasticizer, hereafter called modified single-base (MSB) formulation;
2) A single-base propellant containing NC and an energetic plasticizer;
3) A double-base propellant containing NC, nitroguanidine (NQ), and an energetic plasticizer;
4) A double-base propellant made of NC, cellulose acetate buryrate (CAB), and HMX, hereafter called L320;
5) A double-base propellant containing NC, CAB, HMX, an energetic plasticizer, and an ETPE, hereafter called high-energy low-vulnerability ammunition (HELOVA);
6) A triple-base propellant containing on NC, NQ, and an ETPE.

Because of numerous technical hurdles, described elsewhere [32, 43, 59], only options 1 and 4 were preselected for further development. Their mechanical and ballistic performance and environmental and IM properties are discussed in the following sections.

4.5.2.1 Performance

The conformity of MSB and L320 to the performance prerequisite was verified using formulation quickness, force, and linear burning rate [60]. The performance of a propellant is measured by its burning rate as well as its quickness, which is defined in STANAG 4115 (NATO website, [61]) as the pressurization rate, by its force, which corresponds to the maximum pressure applied at a specific loading charge, and by its vivacity, which is the ratio of quickness to force.

The quickness and force of MSB and L320 are provided in Table 4.5 relative to that of M1-0.025. To find a suitable balance between pressure, burning rate, ignition delay, and firing residues, several shapes, sizes, and webs were studied for L320. Both propellants were considered acceptable for application in 105-mm

Table 4.5 Gun propellant performance at 21 °C and mechanical properties, relative to reference formulation M1-0.025.

Formulation	Relative Young's modulus (%)	Relative quickness (%)	Relative force (%)	Vivacity (%)	Linear burning rate (%)
M1	100	100	100	100	100
MSB	123	130	107	122	105
L320	98	76–117	104	73–112	NA

NA = not available.

artillery munitions. In addition, the range of the L320-filled 105-mm munition was significantly increased.

4.5.2.2 Modular Charges

The propellant charges of a 105-mm artillery munition are currently made of seven different types of bags, each containing a different charge weight; and one bag contains lead, a decoppering agent. In addition, the formulation of the first bag is made of a different shape, size, and web than that in the six other bags. The range of the munition is fine-tuned in the field by removing the appropriate number of bags from the cartridge. The excess propellant bags used to be burnt on the ground immediately after artillery exercises, which produced a significant amount of toxic and carcinogenic contaminants.

The concept of modular charges evaluated within this project consists in building identical charge weight modules, which would be incorporated in the 105-mm cartridge during the firing event. The range would then be fine-tuned using an appropriate number of modules. The net advantages of this concept are that there is no module to burn after each training exercise, and that the deployment logistics is simpler and less expensive, because there is no need to carry supplementary bags that not only may never have to be used in firing but also that would need to be burnt on site. Similar modular charges already exist for the 155-mm artillery munition (M777), and are under study for the 105-mm calibre, albeit with the M1 propellant. The results indicate that the concept of modular charge could be very easily developed with L320.

4.5.2.3 IM Properties

Early batches of propellant formulations with a somewhat different composition than the final one were used for IM tests. Thus, a single-base propellant formulation containing an energetic plasticizer (called MM1) was tested instead of MSB, and HELOVA (option 5) was tested instead of L320. Due to time and budget constraints, those IM tests could not be performed again. The results obtained are believed to be representative of the worst-case scenario, because MSB and L320 did not contain an energetic plasticizer, contrary to MM1 and HELOVA. This hypothesis is supported by hot fragment conductive ignition (HFCI) results for LOVA-HMX, which were better than HELOVA's. The results are summarized in Table 4.6.

Table 4.6 IM tests results for the propellant candidates and reference formulation [49, 62].

Formulation	BI	SCJ	SCO	HFCI[a] (°C)
M1	V	II	IV–V	296–476
MM1	V	II	IV–V	324–500
HELOVA	V	II	IV–V	325–450
LOVA-HMX	NA	NA	NA	350–575

NR = no reaction; NA = not available; Red = fail; Green = pass.
a) For ball weight between 0.24 and 5.60 g.

All formulations passed the BI test, but MM1 and HELOVA demonstrated a less violent reaction than M1. For the SCJ test, the gun propellants reacted similarly with a partial detonation (Type II). As for the IM explosive tests, this level of reaction was expected, considering the severity of the threat. No propellant formulations passed the SCO test: all the propellants candidates produced violent burning reactions similar to that of M1 at comparable temperatures (Type IV–V). The only exception was HELOVA, which was able to resist slightly higher temperatures.

4.5.2.4 Fate, Transport, and Toxicity

M1, MSB, and L320 were very resistant to dissolution [58]. Dissolution tests led to the following order: L320 > M1 > MSB. None of the formulations was expected to dissolve completely. The order of dissolution seems to be related to the proportion of NC in the formulation, a higher NC content leading to a lower lixiviation. It was hypothesized that NC may swell in water and hence slow down the diffusion of the chemicals or act as potential adsorbent for the soluble components via hydrogen bonding.

As indicated in Table 4.7, the ingredients leaching from the propellant formulations were DNT and DPA from M1, HMX from L320, and the plasticizer from MSB and L320. As a matter of fact, up to 15% DNT and 1% DPA were released from M1 after 56 days of contact with water with dripping water, as compared with a little more than one third of the HMX content. The plasticizer is very soluble in water and tends to leach easily from the formulation.

The batch of MSB formulation used for the transport and toxicity tests was stabilized with akardite (AK) and coated with methyl centralite, a deterrent. These compounds were detected at low concentration in the soil leachates or in the soil interstitial water samples.

As indicated in Table 4.7, DPA is the most hydrophobic chemical amongst the five and it exhibits a moderate potential for bioaccumulation; if released in the environment, it is expected to adsorb significantly on soil organic matter. The four other chemicals were found to be less hydrophobic and to exhibit low potential for bioaccumulation. If released in soil, they should sorb partially but in lesser extent than DPA; AK and HMX should be the most mobile amongst them.

Table 4.7 Transport properties and toxicity of M1, MSB, and L320.

Formulation	Released ingredients	Terrestrial toxicity[a]	Sediment toxicity[a]	Aquatic toxicity[a]
M1	2,4-DNT, DPA	+	++	ND
MSB	Plasticizer, AK, methyl centralite	+	ND	++
L320	HMX, Plasticizer	ND	ND	ND

a) The number of + signs is related to the number of tests for which deleterious effects were observed. A total of eight different toxicity tests were performed.

The M1 reference propellant formulation inhibited ryegrass and benthic amphipod growth. The M1 reference formulation, which contains 2,4-DNT, was expected to be toxic to earthworms. The non-toxic effect of 2,4-DNT may be explained by its low bioavailability in the soil interstitial water, which can be explained by the presence of NC on which 2,4-DNT is imbedded.

MSB showed toxic effects to the fresh water algae and duckweed growth, and to earthworm's avoidance behaviour. The source of MSB toxicity is possibly related to the presence of AK and methyl centralite in the soil leachates or in the soil interstitial water samples. It is therefore recommended to replace these compounds by greener ingredients in future batches of MSB.

L320 was not toxic to any of the receptors. Based on these preliminary results, no clear conclusion can be drawn to determine which of the MSB or M1 formulation is more toxic, since the toxic response varied among the selected aquatic, terrestrial, and benthic organisms. The following toxicity gradient can be established. The order of toxicity (least to highest) is LOVA < MSB ≤ M1. However, if AK and methyl centralite are replaced in future MSB batches, the toxicity of the formulation could be further reduced.

4.5.3 Field Demonstration

4.5.3.1 Final Selection

For the continuation of the program, only one explosive formulation and only propellant formulation were selected, based on the previously chosen criteria of performance, cost, environmental impact, and technical feasibility.

For propellant formulations, MM1 and LOVA-HMX outperformed the reference formulation. LOVA-HMX surpassed MM1 for all criteria except cost and technical feasibility, for which there was equality. The LOVA-HMX propellant was thus chosen to pursue the large-scale testing.

For explosive formulations, CX-85 and GIM globally performed better than Comp B. Each candidate is unique and has its strong points: the global environmental properties and performance of the GIM were slightly better than those of CX-85, but the latter slightly exceeded GIM with regard to the IM properties of the formulation. The cost and technical feasibility of Comp B were hard to surpass. Overall, the GIM was deemed a better fit to the project scope and was chosen to pursue the large-scale testing phase of the program.

4.5.3.2 Gun Testing

The goals of this demonstration consisted in (1) testing the functionality of the GIM explosive and the uniformity of its detonation velocity. In addition, the propellant was tested to (i) assess the best configuration, (ii) optimize the low-pressure zones, (iii) verify the zones 1 and 7 characteristics, (iv) the temperature sensitivity, (v) the uniformity, (vi) the reduction or elimination of unburnt residues at lower zones, and (vii) test the modular charges concept. Figure 4.2 shows the instrumented LG1 MKII 105-mm Howitzer employed by the Canadian Armed Forces (CAF) for direct and indirect fire that was used for the demonstration. A total of 65 shots were fired.

Figure 4.2 A 105-mm Howitzer gun used for demonstration.

The results clearly demonstrated that L320 was superior to the current propellant formulation and that modular charges could be very easily developed and implemented with the 105-mm munition [60]. In addition, as shown in Figure 4.3, all projectiles were successfully ignited by the C32A1 fuse. The detonation velocity and pressure were as expected, and similar to that of current 105-mm M1 cartridges [60].

Fragmentation testing using the arena test was also conducted to determine the number and weight distribution, as well as the velocity and spatial distribution of fragments produced upon detonation.

4.5.3.3 Detonation Residues

GIM explosive residues produced during live-fire and BIP operations were assessed by detonating six GIM-filled 105-mm projectiles on ice blocks at DRDC Valcartier experimental test site and collecting the residues on snow. Results,

Figure 4.3 GIM detonation.

Table 4.8 Detonation residues of live-fire and blow-in-place of GIM- and Comp B-filled 105-mm rounds.

	Live-fire		BIP	
	Munition outcome (%)	Residue	Munition outcome (%)	Residue
Comp B[a]	0.000 007	RDX	0.0003	RDX
GIM[b]	0.0003	HMX	0.1	HMX

a) [39].
b) [63].

provided in Table 4.8, were compared to those of Comp-B filled 105-mm munition. The munition's outcomes defined in Table 4.8 refer to the percentage of EM deposited at the detonation point as unburnt post-detonation residue, relative to the original mass in the round, post-detonation.

As expected for most IM munitions, the detonation residues of GIM-filled rounds are one to two orders of magnitude higher than Comp B's. However, GIM-filled rounds performed extremely well compared to other commercially available IM formulations, for which deposition rates were one to two orders of magnitude higher [26, 27]. The lower detonation efficiency of IM munitions is thought to be due either to less sensitive energetic ingredients or to the use of standard high explosives initiators, unmodified for IM munitions. Additional R&D is needed in this area before arriving at a definite conclusion.

4.5.4 Life-Cycle Analysis

A CEA was used to estimate the green munitions' incremental economic costs, based on cost differences between green and conventional munitions [64]. This methodology was preferred to a full cost–benefit analysis because the feasibility of measuring all of the project's benefits (e.g. the value of training) was deemed low. In collaboration with subject matter experts, relevant cost categories were identified during the whole life cycle of the munition, from its manufacture to its disposal either by live-fire or by demilitarization. Obtaining all the necessary data proved to be a colossal challenge, because the information was often missing, partial, or very complex. For example, the RTA maintenance scenario may be as simple as performing a surface clearance to avoid UXO, and cutting the bushes on a flat area, or as complex as performing an in-depth clearance, discarding the UXO by BIP, and cutting trees in a wooded steep area. Despite intensive research efforts, it was not achievable to obtain data for some cost categories. Therefore, simulated data obtained from a hypothetical training installation and realistic baseline scenarios were used for the following categories:

- Liability;
- Remediation of an impact area or firing position;
- Conception of the munitions;

- Manufacturing cost of each unit (shell, propelling charge, fuze, etc.);
- Demilitarization;
- Initial investment (e.g. PBX plant).

Results demonstrated that the mean potential savings of using greener ammo is of several millions of dollars exclusively for the Canadian artillery ranges. The status quo would thus be more expensive due to environmental hazards. Those results are only valid for artillery ranges; drawing conclusions about other types of ranges would involve repeating the same measurements at all RTAs.

4.5.5 Summary

The technology demonstrator project RIGHTTRAC proved that it was possible to develop greener munitions that will perform better than current munitions and that will help ease the environmental pressures on military RTAs. The test vehicle was a 105-mm M1 artillery round, but the technology was meant to be transferable to other calibres. The main outcomes of this project consist of a green and IM main explosive charge, and a greener, modular, extended-range and IM main propellant charge. Each of these outcomes can be exploited separately or as a whole, either for Army, Air Force, or Navy munitions.

4.6 New Enhanced and Green Plastic Explosive for Demolition and Ordnance Disposal

For more than 70 years, Canada as well as many NATO nations have been using large quantities of the plastic explosive Composition C4 for various operational needs. C4 is mostly used as a demolition explosive by engineering units and for the BIP destruction of unexploded ordnances by explosive ordnance disposal (EOD) teams. C4 formulation is mainly based on RDX.

Unexpectedly, even though C4 is a small munition item when compared to large-calibre rounds, C4 blocks were identified as an important source of RDX in Canadian RTAs during the deposition rates studies conducted on C4 [28, 29, 65]. This therefore implies large amounts of potential deposition in active use. C4 leads to dispersion of milligram quantity of RDX per block, which might appear small, but when considering the amount of C4 blocks used per year in RTAs, hundreds of grams of RDX could be dispersed in specific RTAs on a yearly basis. Therefore, alternative options were studied to replace C4 in the Canadian inventory with a greener RDX-free formulation.

PETN and HMX were considered by Canada as potential replacement for RDX-based C4 in plastic explosives. Both are less toxic and much less soluble than RDX. Formulations were obtained either from the industry or from a research centre, under the auspice of the Technical Cooperation Programme (TTCP), panel WPN TP4.

Three types of tests were carried out to evaluate the replacement candidates; performance tests, as options, must at least be as performant as C4; deposition rate tests to measure the quantity of residues post-detonation, as well as fate, transport, and toxicity tests to understand how fast and how wide post-detonation residues would move towards receptors and how toxic they would be. The performance and deposition tests were conducted internally at DRDC, while the fate, transport, and toxicity tests were conducted by the National Research Council Canada (CNRC), an institute that has a strong background in the fate, transport, and toxicity of EM residues [37, 38, 66]. The performance tests conducted on DM12 were described earlier on and are not detailed here [67]. The deposition rate tests were conducted in accordance with the methodology described in Chapter 3 of this book.

4.6.1 PETN Option

A PETN-based formulation was used by Germany as the country's sole plastic explosive for years. These formulations, referred to as Seismoplast or DM-type explosives (DM12 or DM52), were made by the Dynamit Nobel Defence GmbH Company. Interestingly, the CAF extensively used PETN-based formulations in the mid-1980s, when they encountered a shortage of C4 stock. Therefore, Canadian EODs have worked with PETN-based plastic explosives in the past and it did fulfil their needs. Moreover, a large quantity of the commercial DM12 formulation that was used for conducting the tests was still in DRDC's inventory. However, it was thought that PETN-based formulations would be more sensitive to impact and friction than C4. But no incidents or accidents were ever reported by either NATO Munitions Safety Information Analysis Center (MSIAC) or the German Armed Forces Technical Center for Weapons and Ammunition, despite its wide use for many years by the German troops.

4.6.1.1 Performance

The performance tests conducted on DM12 were impact and friction sensitivity tests, shock sensitivity test, electrostatic discharge test, and Cheetah calculations for performance predictions, bulk density, detonation velocity, plate dent test, and handling properties [67]. As predicted, the sensitivity testing of DM12 showed a slighter higher sensitivity than C4, while still within acceptable limits. For the performance, the first step was to undertake Cheetah calculations in an attempt to predict the performance of both explosives. Since PETN has a slightly lower density than RDX and there is somewhat less PETN in DM12 than RDX in Composition C4, the thermodynamic code predicted a lower detonation velocity and a lower detonation pressure for DM12. As stated, it was deemed acceptable by the end users in the past when PETN-based plastic explosive was used by the CAF. The density of an explosive is an important property that relates directly to its performance. The density of both compositions was measured using a few different methods and the bulk densities of both explosives were within the acceptable parameters. The detonation velocity was measured using ionization pins on both bulk samples and on test tubes, in which the compositions were hand-pressed to reach a higher density. The measurements confirmed a somewhat lower detonation velocity for DM12 compared to Composition C4, again still within acceptable limits. The handling properties of both products at low and high temperatures were evaluated, both with qualitative manual handling tests and with compression tests. The handling tests showed a better performance at cold and hot temperatures of DM12, probably because of the silicon oil. The results from these tests showed that DM12 would satisfy the minimum performance requirements.

4.6.1.2 Deposition Rate

To assess how much PETN would be released in the detonation of DM12 blocks, a deposition study was conducted to determine its deposition rate. The details were published in Thiboutot et al. [68] and are therefore not described here. DM12 blocks were placed over pristine snow on ice blocks and detonators were placed in their centre, such as would be done by EOD personnel (Figure 4.4).

Seven blocks were detonated in seven locations in order to get statistically representative results. The methodology described in Chapter 3 of this book was applied to obtain the mass of PETN deposited and its relative deposition rate. No PETN was detected in any of the seven plumes; and using half of the analytical detection limit, it was determined that the deposition rate of PETN was smaller than $1\% \times 10^{-7}\%$ [68]. This is a very low deposition rate, the lowest ever measured for any explosive item, and it is four orders of magnitude smaller than RDX deposition rate from C4.

4.6.1.3 Fate, Transport, and Toxicity

The fate, transport, and toxicity of DM-12 and pure PETN were studied by the CNRC Energy Mine and Environment (EME) Laboratory. They measured PETN solubility and dissolution rates, both in batch and dripping experiments. They also studied PETN transport in soil by determining its partition coefficient (K_d), to estimate the extent to which PETN would be sorbed by soil and thus retarded from migrating of site. It is one of the most critical parameters to evaluate the

Figure 4.4 Set-up for DM12 detonation and deposition rate measurement.

migration potential of a given chemical, and it was estimated using two types of soils representative of DND training range soil types. Another critical parameter studied was the biotic and abiotic degradability of PETN in the environment. This indicates how long PETN would persist in the environment, but, more importantly, what the degradation by-products fate and toxicitiy would be. Finally, the ecotoxicological profile of PETN was also assessed, using a battery of tests based on various ecological receptors, representative of receptors that would be found on military training ranges. Their findings [69, 70] showed a pretty low PETN water solubility of 1.55 mg/l at 25 °C, supporting pervious data published in the literature [71, 72]. Based on its quasi-insolubility, only very small amounts of PETN should dissolve following its release in nature. In connection with the PETN dissolution rate, both batch and dripping experiments showed that less than 0.07% of PETN was released after two months of exposure at 30 °C. These data correspond to the least soluble and slowest dissolution rate ever measured by CNRC's team. This means that the exposure of even small chunks of DM12 to water in the environment would lead to extremely low levels of contamination by PETN. In conclusion, the study on pure PETN and DM12 fate and transport properties demonstrated that PETN shows favourable fate, toxicity, and transport characteristics as it has a very small water solubility, a low toxicity to ecological receptors and it degrades very quickly in the environment to benign by-products. PETN is thus much less toxic, bioavailable, and problematic than RDX. Based on the deposition rate and fate, toxicity, and transport properties measured, the PETN-based plastic explosive was seen as a good option to replace C4 as the deposition rate of PETN is very low, and whatever traces of PETN deposited are not highly water soluble, not highly toxic, and would rapidly degrade to non-toxic degradation products.

4.6.2 HMX Option

The rationale for selecting an HMX option to replace C4 was the following: HMX is a nitramine that has a molecular structure similar to RDX, while it is much less soluble and toxic. EPA drinking water threshold for RDX is 2 ppb, while that of HMX threshold is 400 ppb, or 200 times less stringent [19]. The water solubility of HMX is also 10 times lower than that for RDX (Table 4.1). Therefore, using HMX instead of RDX would reduce the associated risk by at least three orders of magnitude, as the impact of a contaminant is equivalent to its effect, exposure, and fate. Another advantage of an HMX-based plastic explosive is that it would be very similar to C4 for the user community; and HMX being more powerful than RDX, its performance was expected to be better than C4. One drawback of HMX is that it very often contains a certain percentage of RDX, as a production by-product. However, by specifying that an RDX-free HMX was required, it was possible to get very low levels of RDX in HMX from the industry, for a slightly higher cost.

There was no commercially available HMX-based plastic explosive in the market at the time this project was undertaken. However, two types of HMX-based plastic explosive were obtained, one through the reopening of a production line in Norway of an HMX-based paste plastic explosive and one experimental

HMX-based plastic explosive from the US Army ARDEC Explosives Development Branch, Picatinny Arsenal, Picatinny, New Jersey. The Norwegian option, DPX-8116, was based on a polyisobutylene binder of the same type as the C4 binder, while the ARDEC option, PAX-52, was based on a silicon binder, such as DM12.

For the Norwegian option, a contract was awarded to the industry to prepare HMX-based plastic explosive blocks. The Chemring Nobel Company, High Energy Materials Department, located in Norway used to produce a plastic explosive named NM-91 that was very similar to C4 with the exception of being based on HMX. A paste made of a newer version of low-RDX-content HMX-based plastic explosive was ordered from them through the General Dynamics Ordnance and Tactical Systems, Canada. The properties of the paste, such as its density, sensitivity to impact, friction, and ease of processing, were evaluated, and then processed by extrusion. Blocks of HMX-based plastic explosive were then delivered to DRDC for testing.

The fate, transport, and toxicity of both HMX-based formulations were not assessed, as they already had been extensively studied [38] and based on the option analysis rationale, as described earlier. The deposition rates of both HMX-based plastic explosives were tested and the PAX-52 option showed a three times lower deposition rate than that of HMX. Moreover, the PAX-52 handling properties were very good over the range of operational temperatures, similar to those of DM12, whereas the DPX-8116 option showed very poor rheological and handling properties. Both silicon-based formulations DM12 and PAX-52, produced better handling properties than C4.

4.6.3 Summary

A non-negligible environmental risk related to the use C4 in live-fire training ranges was identified. Three greener options were studied, one based on PETN and two based on HMX. The data obtained on the PETN-based option, DM12, showed that it could represent a safe and environmentally friendly option to replace C4 with satisfactory performance. Out of the two HMX-based options, PAX-52 was selected as the preferred candidate for replacing C4 in the Canadian inventory. Therefore, C4 shall be replaced in the future by either DM12 or PAX-52, based on a final selection. It was important for Canada to have at least two options for replacing C4 to enhance the flexibility in case of non-availability of one of the options. This section represents a general holistic approach for the identification deployment of greener and performant plastic explosives.

4.7 Conclusions

One of the safest approaches for reaching military sustainability is the development of greener munitions that even if dispersed in the environment, would not lead to adverse environmental or human health impacts. This approach is seen as a better avenue than conceiving implementing site-specific and expensive military RTA remediation or mitigation methods, because of its universal applicability anywhere in domestic and overseas military activities. However, it is clear that greener munitions cannot be developed at the expense of performance and insensitivity, which are critical for military users. The opposite is also true: Munition development should not be performed at the expense of environment and health, whenever possible.

The development of greener munitions and weapons systems start, of course, from the ingredients themselves that should not be bioavailable or toxic in case of accidental receptors exposure. However, greener molecules themselves are not the Holy Grail; the greener munitions and weapons systems are the goal to reach. Consequently, a holistic approach is required to evaluate the potential of developing greener munitions while also including other parameters, such as performance, insensitivity, demilitarization, ingredients compatibility and stability, formulation processing and properties, ageing, system suitability, and so on. In this regard, the integrated approach described here is a model for developing greener munitions.

In addition, the integrated approach successfully demonstrated that greener munitions are far from being incompatible with performance and cost. The evaluation of the environmental impacts at the beginning of a munition development cycle simultaneously with their performance and IM properties has the benefit of avoiding dedicating considerable efforts on the development of formulations that could be discarded at the end of the development cycle due to noxious environmental impacts. Such a holistic approach can only be adopted and promoted by communities working in close collaboration with each other. Efforts should be dedicated to improve the communication between the munition developer world, as well as the IM and environmental impacts experts to avoid huge investments in weapons development that would be lost afterward, due to the too high risk in their deployment.

Several other challenges are foreseen in the munitions area. For example, nitroglycerin, chlorate, and strontium have been placed on the US EPA Contaminant Candidate List 3 (USEPA website). Contaminants on this list may require regulation under the Safe Drinking Water Act, because of their potential to reach public water systems. NG is commonly used in double-based propellant, chlorate as a replacement of perchlorates in some pyrotechnic applications, and strontium in illuminating flares and tracers. Another emerging contaminant would be the introduction of nanoparticles in energetic formulations, which might lead to unknown adverse environmental impacts. The replacement of these ingredients should be seen as opportunities to apply an integrated approach, such as the one described here, for greener munitions development.

The end result is that military personnel will be able to train and fight with munitions having comparable or better properties than current munitions, with the added benefit of decreasing the environmental pressure and the health hazards on soldiers, sailors, or airmen. The technologies developed under RIGHTTRAC and the RDX-free C4 will contribute to sustain military training and maintain troop readiness by minimizing long-term environmental impacts and provide a model for others to follow and develop to meet future needs.

References

1 Jenkins, T.F., Hewitt, A.D., Grant, C.L. et al. (2006). Identity and distribution of residues of energetic compounds at army live-fire training ranges. *Chemosphere* 63: 1280–1290.
2 Pennington, J.C., Jenkins, T.F., Ampleman, G. et al. (2006). Distribution and Fate of Energetics on DoD Test and Training Ranges. *Final Rep. ERDC TR-06-13*, Cold Regions Research and Engineering Laboratory, Hanover, NH.
3 Thiboutot, S., Ampleman, G., Brochu, S. et al. (2012). Environmental characterization of military training ranges for munitions-related contaminants. *Int. J. Energetic Mater. Chem. Propul.* 11 (1): 17–57.
4 Ampleman, G., Thiboutot, S., Cinq-Mars, A. et al. (2013). Development of a Small Arms Bullet Catcher. *DRDC Valcartier TR-400*, Defence Research and Development Canada, Valcartier, QC, Canada.
5 Finland Ministry of the Environment (2014). Best available techniques (BAT) management of the environmental impact of shooting ranges. In: *The Finnish Environment*, vol. 4 (ed. S. Kajander and A. Parri).
6 Interstate Technology and Regulatory Council (ITRC) (2005). Environmental management at operating outdoor small arms firing ranges. SMART-2. Washington, DC: Interstate Technology and Regulatory Council, Small Arms Firing Range Team. https://www.itrcweb.org (accessed 14 May 2018).
7 Brinck, T. (2014). *Green Energetic Materials*, 1ee. John Wiley & Sons Ltd.
8 Environmental Protection Agency (USEPA) (1997). Region 1, Administrative Order. EPA Docket Nos.: SDWA I-97-1019 and SDWA I-97-1030.
9 Environmental Protection Agency (USEPA) (2014). Technical fact sheet – tungsten. https://www.epa.gov/sites/production/files/2014-03/documents/ffrrofactsheet_contaminant_tungsten_january2014_final.pdf (accessed 14 May 2018).
10 Agency for Toxic Substances and Disease Registry (ATSDR) (2005). Toxicological profile for tungsten. https://www.atsdr.cdc.gov/toxprofiles/tp.asp?id=806&tid=157 (accessed 14 May 2018).
11 Clausen, J. and Korte, N. (2009). Environmental fate of tungsten from military use. *Sci. Total Environ.* 407: 2887–2893.
12 Association of State and Territorial Solid Waste Management Officials (ASTSWMO) (2011). Tungsten issues paper. http://astswmo.org/files/policies/Federal_Facilities/TUNGSTEN_FINAL_120208.pdf (accessed 14 May 2018).
13 Voie, Ø., Borander, A.-K., Sikkeland, L.I.B. et al. (2014). Health effects after firing small arms comparing leaded and unleaded ammunition. *Inhalation Toxicol.* 26 (14): 873–879.
14 Walsh, M.R., Walsh, M.E., Ramsey, C.A. et al. (2013). Perchlorate contamination from the detonation of insensitive high-explosive rounds. *J. Hazard. Mater.* 262: 228–233.
15 Walsh, M.R., Walsh, M.E., Taylor, S. et al. (2013). Characterization of PAX-21 insensitive munition detonation residues. *Propellants Explos. Pyrotech.* 38 (3): 399–409.
16 Sabatini, J. (2014). Advances toward the development of "Green" pyrotechnics, Chapter 4. In: *Green Energetic Materials* (ed. T. Brinck). John Wiley & Sons Ltd.

17 Interstate Technology and Regulatory Council (ITRC) (2005). Perchlorate: overview of issues, status, and remedial options. Washington, DC: Interstate Technology and Regulatory Council. https://www.itrcweb.org (accessed 14 May 2018).
18 Government of Canada (GC) (2012). Health Canada, Perchlorate and Human Health.
19 Environmental Protection Agency (USEPA) (2012). 2012 Edition of the Drinking Water Standards and Health Advisories. *EPA 822-S-12-001*.
20 European Union (EU) (2018). Evaluation, authorisation and restriction of chemicals (REACH) – substances of very high concern. https://echa.europa.eu/regulations/reach/substance-identity (accessed 14 May 2018).
21 Director Land Environment (DLE) (2010). Army bases range and training area characterization. Department of National Defence, Government of Canada.
22 Taylor, S., Dontsova, K., Bigl, S. et al. (2012). Dissolution Rate of Propellant Energetics from Nitrocellulose Matrices. *ERDC/CRREL TR-12-9*, Cold Regions Research and Engineering Laboratory, Hanover, NH.
23 Thiboutot, S., Ampleman, G., Kervarec, M. et al. (2011). Development of a Demilitarisation Table for the Safe Burning of Excess Artillery Propellant Charge Bags. *RDDC Valcartier TR-2010-254*, Defence Research and Development Canada, Valcartier, QC, Canada.
24 Thiboutot, S., Ampleman, G., Kervarec, M. et al. (2011). Fix and mobile demilitarisation units for the burning of excess artillery propellant charge bags, Chapter 4. In: *Characterization and Fate of Gun and Rocket Propellant Residues on Testing and Training Ranges: Final Report, ERDC/CRREL Rep. TR-11-13*, 77–128. Hanover, NH: Cold Regions Research and Engineering Laboratory.
25 Thiboutot, S., Ampleman, G., Brochu, S. et al. (2012). Surface Soils Sampling for Munition Residues in Military Live-fire Training Ranges. *DRDC Valcartier TR 2011-447*, Defence Research and Development Canada, Valcartier, QC, Canada.
26 Walsh, M.R., Walsh, M.E., Ramsey, C.A. et al. (2015). Energetics residues deposition from training with large caliber weapon systems. *2nd European Conference on Defence and the Environment*, Helsinki, Finland (9–10 June 2015).
27 Walsh, M.E., Walsh, M.R., Ramsey, C.A. et al. (2015). Collection, processing, and analytical methods for the measurement of post-detonation residues from large caliber ammunition. *2nd European Conference on Defence and the Environment*, Helsinki, Finland (9–10 June 2015).
28 Diaz, E. and Thiboutot, S. (2016). Deposition rates from blow-in-place of different donor charges: comparison of composition C-4 and shaped charges. *Propellants Explos. Pyrotech.* 42 (1): 90–97.
29 Walsh, M.R. (2006). Explosives Residues Resulting from the Detonation of Common Military Munitions: 2002–2006. *ERDC/CRREL Technical Rep. TR-07-2*, Cold Regions Research and Engineering Laboratory, Hanover, NH.
30 Walsh, M.R., Walsh, M.E., Poulin, I. et al. (2011). Energetic residues from the detonation of common US ordnance. *Int. J. Energetic Mater. Chem. Propul.* 10 (2): 169–186.
31 Lewis, J., Martel, R., Trépanier, L. et al. (2009). Quantifying the transport of energetic materials in unsaturated sediments from cracked unexploded ordnance. *J. Environ. Qual.* 38: 1–8.

32 Chavez, J. (2014). Étude du mécanisme de corrosion des munitions, Rapport B-0007995-1-MC-0001-00, Contract DVRD2013, Dessau.

33 Thiboutot, S., Ampleman, G., Gagnon, A. et al. (2010). Nitroglycerin fate at a former antitank range firing position. In: *Characterization and Fate of Gun and Rocket Propellant Residues on Testing and Training Ranges: Interim Report 2, ERDC/CRREL Report TR-10-13*, Chapter 5, 109–144.

34 Walsh, M.R., Walsh, M.E., Thiboutot, S. et al. (2012). Munitions propellants residue deposition rates on military training ranges. *Propellants Explos. Pyrotech.* 37: 393–406.

35 Environmental Protection Agency (USEPA) (2018). Green chemistry. https://www.epa.gov/greenchemistry (accessed 14 May 2018).

36 Monteil-Rivera, F., Halasz, A., Manno, D. et al. (2009). Fate of CL-20 in sandy soils: degradation products as potential markers of natural attenuation. *J. Environ. Pollut.* 157 (1): 77–85.

37 Perreault, N., Monteil-Rivera, F., Halasz, A. and Dodard, S. (2017). Environmental Fate and Ecological Impact of Energetic Chemicals, Twenty-year Study. *Summary Rep. 1996-2016, NRC-EME-55873*, National Research Council Canada.

38 Sunahara, G.I., Lotufo, G., Kuperman, R.G., and Hawari, J. (2009). *Ecotoxicology of Explosives*. CRC Press.

39 Hewitt, A.D., Jenkins, T.F., Ranney, T.A. et al. (2003). Estimates for Explosives Residue from the Detonation of Army Munitions. *ERDC/CRREL-TR-03-16*, Cold Regions Research and Engineering Laboratory, Hanover, NH.

40 Jenkins, T.F., Bartolini, C. and Ranney, T.A. (2003). Stability of CL-20, TNAZ, HMX, RDX, NG, and PETN in Moist, Unsaturated Soil. *ERDC/CRREL-TR-03-7*, Cold Regions Research and Engineering Laboratory, Hanover, NH.

41 Puhlmann, S. (2006). Assessment of the DM12B1 Demolition Charge – Environmental Impact Analysis. *Rep. C/E210/68798/0000*, Bunderswerh Research Institue for materials, Fuel and Lubricants.

42 Meyer, R., Köhler, J., and Homburg, A. (2007). *Explosives*, 6ee. Wiley.

43 Brochu, S., Brassard, M., Ampleman, G. et al. (2011). Development of high performance, greener and low vulnerability munitions – revolutionary insensitive, green and healthier training technology with reduced adverse contamination (RIGHTTRAC). In: *NATO AVT-177, Munitions and Propellant Disposal and its Impact on the Environment*, Edinburgh, UK (17–20 October 2011).

44 Brochu, S., Brassard, M., Thiboutot, S. et al. (2014). Towards high performance, greener and low vulnerability munitions with the RIGHTTRAC technology demonstrator program. *Int. J. Energetic Mater. Chem. Propul.* 13 (1): 7–36.

45 Ampleman, G., Marois, A., and Désilets, S. (2002). Energetic copolyurethane thermoplastic elastomers. Canadian Patent 2,214,729, European Patent Application No.: 0020188.2-2115, US Patent 6,479,614 B1, filed 2 March 2000 and issued 12 November 2002.

46 Ampleman, G., Brousseau, P., Thiboutot, S. et al. (2003). Insensitive melt cast explosive compositions containing energetic thermoplastic elastomers. US Patent 6,562,159, filed 26 June 2001 and issued 13 May 2003.

47 Ampleman, G. (2011). Development of new insensitive and greener explosives and gun propellants. *NATO AVT-177/RSY-027 Symposium, Munition and*

Propellant Disposal and its Impact on the Environment, Edinburgh, UK (17–20 October 2011).

48 Brousseau, P., Brochu, S., Brassard, M. et al. (2010). Revolutionary insensitive, green and healthier training technology with reduced adverse contamination (RIGHTTRAC) technology demonstrator program. *41st International Annual Conference of ICT*, Karlsruhe, Germany (29 June–2 July 2010).

49 Brousseau, P., Brochu, S., Brassard, M. et al. (2010). RIGHTTRAC Technology Demonstration Program: Preliminary IM Tests. In: *Insensitive Munitions and Energetic Materials Technical Symposium*, Munich, Germany (11–14 October 2010).

50 Brochu, S., Williams, L.R., Johnson, M.S. et al. (2013). Assessing the Potential Environmental and Human Health Consequences of Energetic Materials: A phased Approach, *TTCP WPN TP-4 CP 4-42, Final Rep., TTCP TR-WPN-TP04-15-2014*.

51 Government of Canada (GC) (2018). List of challenge substances. https://www.canada.ca/en/health-canada/services/chemical-substances.html (accessed 14 May 2018).

52 Government of Canada (GC) (2018). Canadian council of ministers of the environment. http://ceqg-rcqe.ccme.ca/en/index.html (accessed 14 May 2018).

53 Agency for Toxic Substances and Disease Registry (ATSDR) (2018). https://www.atsdr.cdc.gov/ (accessed 14 May 2018).

54 Environment Canada (EC) and (Health Canada) HC (2011). Screening assessment for the challenge, hexanedioic acid, bis(2-ethylhexyl) ester (DEHA). In: *Chemical Abstracts Service Registry Number 103-23-1*.

55 Hawari, J., Monteil-Rivera, F., Radovic, Z., et al. (2009). Environmental Aspects of RIGHTRAC TDP-Green Munitions. *Rep. NRC #49958*, Biotechnology Research Institute of National Research Council, Montréal, QC, Canada.

56 Hawari, J., Radovic, Z., Monteil-Rivera, F., et al. (2011). Environmental Aspects of RIGHTRAC TDP-Green Munitions. *Annual Rep. NRC #53361*, Biotechnology Research Institute of National Research Council, Montréal, QC, Canada.

57 Hawari, J., Monteil-Rivera, F., Radovic, Z., et al. (2010). Environmental Aspects of RIGHTRAC TDP- Green Munitions. *Interim Rep. NRC #50014*, Biotechnology Research Institute of National Research Council, Montréal, QC, Canada.

58 Hawari, J., Radovic-Hrapovic, Z., Monteil-Rivera, F., et al. (2012). Environmental Aspects of RIGHTRAC TDP- Green Munitions. *Annual Rep. NRC #53417*, Biotechnology Research Institute of National Research Council, Montréal, QC, Canada.

59 Petre, C.F., Paquet, F., Brochu, S., and Nicole, C. (2012). Optimization of the mechanical and combustion properties of a new green and insensitive gun propellant using design of experiments. *Int. J. Energetic Mater. Chem. Propul.* 10 (5), 437, 201.

60 Dietrich, S., Éthier, P.-A., Lahaie, N. and Pelletier, P. (2014). Revolutionary Insensitive Green and Healthier Training Technology with Reduced Adverse Contamination (RIGHTTRAC). *Final Rep.*, General Dynamics Ordnance and Tactical Systems, Canada.

61 North Atlantic Treaty Organization (NATO) STANAG 4115 – definition and determination of ballistic properties of gun propellants. https://www.nato.int (accessed 14 May 2018).

62 Petre, C.F., Tanguay, V., Brousseau, P., and Brochu, S. (2011). Use of drop weight and hot fragment conductive ignition tests to characterize new green and insensitive gun propellant, *42nd International Annual Conference of the Fraunhofer ICT*, Karlsruhe, Germany (28 June–1 July 2011).

63 Walsh, M.R, Walsh, M.E., Taylor, S. et al. (2013). Characterization of Residues from the Detonation of Insensitive Munitions. Interim Rep., *SERDP Project ER-2219*, Cold Regions Research and Engineering Laboratory, Hanover, NH.

64 Sokri, A. (2015). Cost Risk Analysis of Green and Insensitive Munitions. *DRDC-RDDC-2015-R056*, DRDC – Centre for Operational Research and Analysis.

65 Thiboutot, S., Ampleman, G., Brochu, S. et al. (2013). Canadian programme on the environmental impacts of munitions – deposition rate of explosives, propellants and IM explosives. In: *Proceedings of the European Conference of Defence and the Environment*, Helsinki, Finland.

66 Sheramata, A.T.W., Halasz, A., Paquet, L. et al. (2001). The fate of the cyclic nitramine explosive RDX in natural soils. *Environ. Sci. Technol.* 35: 1037–1040.

67 Thiboutot, S., Brousseau, P., Ampleman, G. et al. (2015). Alternative options to C4 plastic explosive to mitigate RDX dispersion in ranges and training areas. *NATO STO-RSM-AVT-244, DRDC-RDDC-2015-P123, NATO Unclassified and Partners for Peace*, Finland and Sweden (1–12 October 2015).

68 Thiboutot, S., Brousseau, P., and Ampleman, G. (2015). Deposition of PETN following the detonation of seismoplast plastic explosive. *Propellants Explos. Pyrotech.* 40 (3): 329–332.

69 Hawari, J., Sunahara, G.I., Perreault, N. et al. (2014). *Environmental Fate and Ecological Impact of Emerging Energetic Chemicals (ADN, DNAN and its Amino-Derivatives), PETN, NTO, NQ, FOX-7, and FOX-12 and an Insensitive Formulation*. National Research Council, Energy Mine and Environment.

70 Perreault, N., Halasz, A., Dodard, S. et al. (2016). *Ecological Impact of Emerging Energetic Chemicals (ADN, DNAN and its Amino-Derivatives), PETN, NTO, NQ, FOX-7, and FOX-12 and Insensitive Formulation 2015/2016*. National Research Council.

71 Merrill, E.J. (1965). Solubility of pentaerythritol tetranitrate in water and saline. *J. Pharm. Sci.* 54: 1670–1671.

72 Budavari, S. (ed.) (1989). *Merck Index*, 7068. Rahway, NJ: Royal Society of Chemistry.

5

Pyrotechnics and The Environment

Ranko Vrcelj

Centre for Defence Chemistry, Cranfield University, Defence Academy of the UK, Shrivenham, Swindon, SN6 8LA, UK

5.1 Introduction

When the general public thinks of pyrotechnics, it is normally as part of an exciting celebratory fireworks display. However, pyrotechnic materials and devices are ubiquitous in the modern world with both civilian and military applications, whether it is a distress flare, a smoke screen, or the humble match.

This ubiquity in the modern world means that as with all products, their environmental impact needs to be considered. This impact can be immediate and is surely recognizable to anyone who attends fireworks displays with smoke generated by burning compositions, the associated 'rotten eggs' smell of sulfur-based combustion products, or even from the litter remains after the event. What may be less obvious to many users of pyrotechnics is that many compositions often contain or produce considerable amounts of objectionable or even hazardous chemicals and it is necessary to determine and understand the environmental effect of these materials. This is not necessarily well-understood. In examining these effects and any potential ways to ameliorate them, it is vital that we consider the past and current context of pyrotechnic systems, before looking to the future.

Historically, pyrotechnic compositions have been optimized very simply with respect to their performance. Once a composition was found to work well, it will have been tweaked with additives, such as binders, colourants, etc. to optimize the performance for any given required effect, whether that is heat, light, smoke, sound, or gas (often from a military perspective). A typical example is given in Table 5.1, which examines the different formulations of historic versions of gunpowder – the earliest useful energetic and pyrotechnic known to man.

Gunpowder also gives the first truly environmental statement 'The fog of war' – the smoke and particulates emitted as firing progressed limited visibility and no doubt had an unpleasant effect on the health of those unfortunates in the middle of battle.

Energetic Materials and Munitions: Life Cycle Management, Environmental Impact and Demilitarization,
First Edition. Edited by Adam S. Cumming and Mark S. Johnson.
© 2019 Wiley-VCH Verlag GmbH & Co. KGaA. Published 2019 by Wiley-VCH Verlag GmbH & Co. KGaA.

Table 5.1 Historic gunpowder compositions.

	Explosive	Propellant (small rockets)	
		Germany	United States
Potassium nitrate	75	60	59
Carbon	15	25	31
Sulfur	10	15	10

Along with the development of organic chemistry in the mid-nineteenth century and the splitting of energetic materials categories into explosives, propellants, and pyrotechnics, there also is a splitting of environmental factors. The relatively limited number of explosives and propellants used are almost entirely organic-based, with the exceptions in aluminium and perchlorates in aluminized explosives and high-performance propellants. Pyrotechnics on the other hand, use nearly everything. The list of ingredients is long; metals, non-metals, inorganic salts, organic fuels, inorganic oxyacid salts, metal oxides, organic halogen compounds, natural organic binders, and synthetic organic binders. And this list is by no means exhaustive.

The balance of usage and environmental effect is subtle, particularly for pyrotechnics. They are in every modern weapons system, but at a wide range of different levels, from less than a gram as part of a detonating system, through to the kilogram scale in full pyrotechnic munitions, as opposed to the much larger scale involved for explosives and propellants. So although there is much less use of pyrotechnics, this is countered by the breadth of materials used and the range of applications. Another difference is that the products of both explosives and propellants are designed to create gases, either for propagation of a projectile, or the shock wave of detonation. In a pyrotechnic system, this is often not the case and much solid residue remains after the use of the pyrotechnic.

In examining these polluting effects, it is important to consider where they may arise. Ideally, any product is made and used, without any waste. While it is equally true in the energetics sector, it is clear that there is waste involved and their usage is the deployment of a chemical reaction to cause an effect. Thus, we have to examine these 'wasteful' processes and what the by-products are of using any munition.

The drive for developing green pyrotechnic systems has come primarily from the United States, in particular, with the US Environmental Protection Agency (EPA) establishing guidelines for perchlorate levels in the environment, in addition to the more general concern regarding the levels of heavy metal contamination. These guidelines have encouraged the US Department of Defense (DoD) to fund a broad and wide-ranging set of projects to examine these very issues, in particular through the offices of the Strategic Environmental Research and Development Program (SERDP) and the Environmental Security Technology Certification Program (ESTCP). These have focused on such areas as removal of perchlorates, removal of heavy metals, and a drive to solvent-free manufacture.

The initial results of some of these studies have been more fully explored in the excellent review chapter by Sabatini and in a subsequent review paper on illuminating compounds [1, 2]. Where there have been advances since these reviews, the author will discuss them in the relevant section.

What is of interest is that such a directed set of studies has not yet been fully replicated outside the United States. Within Europe, there are groups actively engaged (e.g. that of Professor Thomas Klapötke, University of Munich) with such research, although all their funding has been derived from the DoD. In the United Kingdom, there is equally little movement, with the only recent study being that of an attempt to move to lead-free primary explosives. There has been some attempt under the auspices of the European Defence Agency (EDA), but this too has not developed as well as it might have.

There is some development in China, but it is yet difficult to see whether this is part of a general 'clean-up' strategy, driven by the Chinese Government/Defence, or aspects of research from interested individuals or research groups. It is likely to be the former; but with the understandable opacity of governments who are unwilling to release their strategy for defence materials, we can only guess.

Although there is little concerted defence effort to move to 'greener' pyrotechnics within Europe, there is a focused and strong drive emanating from the European Union (EU) for a broader control on polluting materials within the member states and this, in place of a Defence sector initiative, is the main driving force behind change within the EU.

5.2 Registration, Evaluation, Authorisation and Restriction of Chemicals (REACH)

The Registration, Evaluation, Authorisation and Restriction of Chemicals (REACH) came into force in 2007 as a central piece of legislation to replace a number of older EU and national directives regarding hazardous chemicals, and the overall approach to REACH is discussed elsewhere in this work.

On the list of substances of very high concern (SVHC), there are currently 43 materials (in the most recent list, published in December 2017) that require authorization and another 138 materials that are candidates for addition to the list (https://echa.europa.eu/regulations/reach/candidate-list-substances-in-articles) and it is expected that many more will follow.

Although there are no secondary explosives on this list, there are lead-based primary explosives, e.g. lead azide and styphnate. However, for pyrotechnics, the case is not so simple. Not only are there a number of materials that have been extensively used as part of pyrotechnic compositions, e.g. lead chromate and potassium dichromate, but also materials that are part of pyrotechnic manufacturing or devices, e.g. trichloroethane (TCE) and dibutyl phthalate (DBP).

Thus, the age-old reliance on heavy-metal-based pyrotechnic compositions and a range of organic solvents is coming to an end and both the materials used within pyrotechnic compositions must change, as do the methods for manufacturing them. For the EU, this has been the main driving force behind changes in the pyrotechnics sector of the defence industries.

There are some possible hiccups on the way. Within the EU, at least the situation is clear; however, for the United Kingdom it is less so, following the 2016 UK referendum result to leave the EU. Negotiations between the United Kingdom and the EU are ongoing and at the time of writing, there is no real clarity as to what any future UK legislation may incorporate, although the UK Government's response to a House of Commons Environmental Audit Committee report has been mutedly favourable [3, 4]. This implies that whatever the outcome of the negotiations, there will be a push to keep some form of centralized chemicals regulation structure and there is some expectation that given the general drive to use more environmentally benign, that the essence of REACH will still be in place after the United Kingdom leaves the EU.

The observations of Sabatini [1] that REACH '...is believed to be the strictest law to date regulating chemicals' are to some degree justified, but we must see this as a positive challenge in two particular directions: to move forwards using new and innovative science to create new materials and to develop strong and well-defined supplier/manufacturer/customer relationships. Indeed, the experience of industry within the United Kingdom is that a knowledgeable and REACH-compliant supplier is worth their weight in gold. But it is also the experience of industry that many customers could be, and, in some cases, still are slow to understand the implications of REACH. Some customers simply ignored (and still ignore) the effects of legislation, not realizing that their manufacturer will have to change, even if they do not consider that they have to. Some national customers (e.g. Switzerland) are very forward-looking and require all the appropriate authorization and documentation prior to placing an order.

The supply of products to a customer necessitates a safety data sheet (SDS) for environmental impact and a statement that no materials in the store are prohibited in REACH has to be supplied.

REACH will have some implications; however, it currently does not affect pyrotechnics as much as it does other sectors, but this does not mean that complacency can set in. There will be further additions to the list of authorized chemicals, but at least the slowness of the additions are partly due to the inevitable slowness with which any large organization works and partly to permit replacements to be found; for example, the search to replace lead azide and lead styphnate in primary explosives is driven by the placing of these two substances on the SVHC.

The development of future environmentally benign pyrotechnic systems is directed by both legislation (e.g. REACH) and economic factors. Along with the efforts of REACH to improve the environmental impact of any given chemical in use, financial viability must come into play. Because much of Europe has been affected by the economic downturn of the past 10 years, funding for new projects and research is inevitably hit hardest, and thus most likely the cause for the lack of future horizon scanning and funding for replacements.

5.3 Qualification

One major issue that remains for the implementation of environmentally benign materials is that of qualification for use. In the United Kingdom, qualification is the rigorous testing that a whole munition must undergo, prior to that munition's use being sanctioned for use by the Ministry of Defence (MOD), analogous to homologation in France and other similar approaches on mainland Europe. These tests are extensive and expensive. This has important implications for the defence industry as opposed to many others in that changing a supplier or manufacturing process requires any particular store to be qualified for use. Whilst this is incorporated into the design of a new store, it requires extra expense in an existing system and often the cost outweighs the interest so that a cost–benefit analysis is needed. The defence sector is notoriously conservative to innovation, as there are issues such as the effect on service life and stability. In addition, there is a long legacy of data for the current range of materials; and creating a completely new composition would require starting performance and characterization data collection again. Thus, any change is only cost-effective for large-volume items or if imposed through other routes.

So, in examining the future of the environment and pyrotechnics, within the EU at least, there must be an examination of current production practices, as well as efforts to replace the objectionable materials they might contain.

5.4 Civilian Studies

Whilst the focus of this chapter is the environmental impact of military pyrotechnics, there is also a broad range of studies of the impact of civilian pyrotechnic systems. However, there is clearly little crossover between military and civilian usage of explosives, this is not the case for civilian/defence crossover in pyrotechnics. Not only are many of the compositions similar, e.g. metal powders and oxidizing agents, but the manufacturing is also similar, even if the quantities are sometimes at different scales. Although the required effects are more limited for civilian pyrotechnics, flash, coloured illumination, etc., the environmental effects are likely to be similar and, if nothing else, more easily and regularly monitored. Thus, this section concentrates on the studies of the environmental impact of civilian pyrotechnics – where a crossover can be seen to be relevant. It should also be remembered that it will most likely be a civilian monitoring of environmental impact that shows any polluting effects, for example, launch of ever more sensitive environmental monitoring systems such as the recent SENTINEL project (https://sentinel.esa.int/web/sentinel/home) or the recent identification of the local environmental damage caused by the chlorosulfonic acid–based smoke obscuration of the Tirpitz in World War II, identified by dendrochronology has recently come to light [5]. In this manner, as environmental concerns become more problematic, there will be fewer places for manufacturers and users to hide, even from historic pollution events.

Civilian studies were originally scant, being based on suggestions that health issues may be related to manufacturing or firework displays [6–9], but without any clear-cut conclusions. However, these types of studies have proliferated over the past 20–25 years, with many aspects being studied, ranging from a study on the long-term (~40 years) effects of aluminosis on workers within the industry [10], through the effects of perchlorates in local water supplies to recent numerous studies on the effect of particulates on the atmosphere. There are also a few reports regarding how to ameliorate the worst of the pollution effects. The recent studies are better placed within the modern context of both pollution and health and safety controls at work, although the effects of pollutants as by-products are most likely unchanged from the early days.

Civilian studies are worthwhile as they are able to place baselines on collectable data, even if the events themselves are not necessarily controllable. For example, the use of fireworks as part of a display or celebration can be controlled by dates and events, the traditional Bonfire Night events in the United Kingdom on (or around) 5 November each year, permits a pre-display baseline to be determined, activity to be monitored, and post-event studies to take place, even if the displays themselves are uncontrolled by the observers [11, 12]. This type of study is occurring all over the world, in particular in regions where fireworks are commonly used in large-scale celebrations, in particular, China and India. Both these countries have started to examine the effects of pollutants on the inhabitants of their countries and many of the reports are cited here.

The studies that have dominated have been those associated with pollution models across the board, generally examining the dispersal of heavy metals, organic and organochlorine by-products of combustion and other noxious gaseous products, e.g. SO_2 and NO_x gases as aerosols and the issues of perchlorate dispersion.

The landmark publications which stirred the community interest and showed the impact that civilian pyrotechnics can have began with the findings of Dyke et al. [13] that during the UK Bonfire Night celebrations, atmospheric dioxins in the area surrounding the activity were enhanced by a factor of at least 4 (dependent on the local conditions), as well as elevation of particulate matter (PM) of the 10-μm size (PM10), although this study could not differentiate between particulates derived from the fireworks themselves, or the accompanying bonfire. A more controlled laboratory-based study of Fleischer et al. [14] showed that some fireworks did evolve enhanced levels of dioxins and dibenzofurans, in particular, those associated with blue light emission. Whilst the work of Lee et al. [15] showed that within a UK context, the dioxin/dibenzofuran cocktail of pollutants can be more readily explained by bonfire burning (as opposed to contributions from fireworks), the most recent study by Schmidt et al. [16] have shown that chlorophenols, chlorobenzenes, and dioxins released by fireworks can contribute up to 10% of the general background. This study is an excellent description of a pollutant finder; and as it is Swiss-based, it removes any extra confusion associated with bonfire contributions to UK studies.

A move from organochlorine materials to airborne metallic particles was first addressed by Moreno et al. [17], examining metal-containing aerosols and highlighted the high PM2.5/10 ratio during transient fireworks events and particularly the enhanced PM2.5 emissions of SO_2, NO, and metals. This was enhanced by the later study of Moreno et al. [18], which showed a significant peak in heavy metal pollutants, in particular Pb. This work was enhanced by the studies of Vecchi et al. [19] who showed significant increases in airborne metal particulates of Sr, Mg, K, Ba, and Cu at PM10 and that of Godri et al. [20] who showed that pyrotechnic-activity-based PM has more effect on health than traffic PM released at the same time.

The work of Steinhauser et al. [21–23] acted as a major focus on the issues of fireworks and pollution. His landmark paper examining pollution in Austria showed that metal particulates can be found in snow, in particular that barium is enhanced by a factor of up to 56 000, compared to the pre-fireworks snow as well as visibly contaminated snow contains related metal particulates, such as strontium, potassium, and iron. As well as expected metal contaminants, radium and arsenic have been shown to be in evidence within civilian fireworks, either as part of illicit purchases, or from raw materials contamination.

A flurry of studies from Malta showed the effects of fireworks not only on the ambient air quality but also on perchlorate-laden dust indoors. The small size of the island and focused celebratory festivals give a strong insight into how fireworks can affect the local atmospheric environment [24, 25], with strong correlations between Al, Ba, Cu, Sr, and Sb and PM10 concentrations showing how the long festival period can significantly contribute to poor air quality in a localized environment. In addition, perchlorate-laden dust remains indoors for a significantly long period of time – leading to possible ingestion and inhalation in a domestic context [26]. Similar spiking of values of perchlorates in ground and surface water was shown by Scheytt et al. [27] in their study of post-fireworks displays in Germany.

Such studies have been extended by work such as that of Joly et al. [28], Crespo et al. [29], Baranyai et al. [30], Remiškar et al. [31], examining heavy metal pollutants, that of Attri et al. [32] and Caballero et al. [33] who examined the role of ozone as pollutant, and finally that of Ten Brink et al. [34] studying the role in diminished visibility that soot and aerosols can play.

5.5 Production

When we consider the environmental impact of pyrotechnics, we have to consider where the impact is centred on, using the Source – Pathway – Receptor model of environmental impact. In a coarse outline, there are the following stages:

- Production
- Production storage
- Transportation to customer
- Customer storage
- Customer transportation
- Customer usage
- End-of-life (transportation and disposal)

Waste or usage occurs at all stages of the life cycle. We may consider the transportation to be environmentally neutral – this is not true from the fossil fuel perspective, but is so from the effects of the pyrotechnics. This does not attempt to minimize the environmental impact of transportation. From a global perspective use of fossil fuels, say, either of heavy goods vehicles or diesel-driven ships will contribute to ongoing climate change, although the levels depend on the distances travelled. For example, a recent calculation was performed for a major civilian firework event in the United Kingdom and it was estimated that there was a greater environmental footprint caused by the exhaust gases in travelling the 30–40 mi (50–65 km) from the storage depot to the place of the display, than contributed from the fireworks display itself.

Storage, whether by the manufacturer or by the customer, can be considered as a single environmental point and to some degree so can usage and disposal. One uses the pyrotechnics for the desired outcome, the other simply destroys it. Nevertheless, the process is the same, the using up of the pyrotechnic. This leaves us with three main focus points: production, storage, and usage/end-of-life disposal which will produce pollutants from pyrotechnics.

These are the three sources that can be utilized for an examination of the Source – Pathway – Receptor model. Some of this approach is discussed in more detail elsewhere in this volume.

5.6 Site Location

First and foremost, the production site production is of paramount importance, as the space available, adjacent neighbours, and geological and geographical features determine the likelihood of any pollution staying within a production site or moving elsewhere. In addition, the history of the site may contribute to environmental issues.

Within the United Kingdom, although production plants were often situated in more remote regions, a combination of limited space and the urbanization of rural areas means that much more attention now needs be paid to the local environment. In addition, the increased value of land (even for a brown site such as an ex-industrial site would be) has led to smaller and smaller production areas as land is sold off for development. Therefore, more stringent waste management systems are required within the United Kingdom and any other country that has a large population density (e.g. The Netherlands). Countries with a lower population density may consider that these waste management systems are not needed; but ultimately if they are developed and implemented now, it will be a boon for the future.

Many, but not all, pyrotechnic manufacturing plants are sited on the remains of older civilian fireworks companies, bringing along with it additional burdens. Regular soil sampling is required to ensure that the legacy of the older sites is understood and managed, and in one case it is known that a previous manufacturer used to bury its waste. While it might be feasible to examine old documentation to find out where exactly previous waste was deposited, it is highly unlikely that there will be any accurate records of what those waste disposal sites contain.

This may also be difficult where records have been lost or never produced due to warfare or regime change.

As urbanization increases, the limits to what can be done on-site also increase, in particular for localized trials areas. While air countermeasures can be tested in a relatively small space, the testing of smokes and obscurants requires a much larger (and thus more expensive) area; and in the case of water-based pyrotechnic munitions, the only suitable site is on the sea itself. Inland facilities do exist, but the encroachment of housing means it will become harder to perform even the earliest and most basic of water-borne tests, prior to trials offshore. Whether the neighbours are a new housing estate, a golf course, or woodland with seasonal hunting occurring, pressure will always arise when a local community impinges on an existing manufacturing site.

The downsizing of production sites is accompanied by the economic considerations of modern manufacturing and to some degree aids any environmental concerns. The concept of 'just-in-time' manufacturing encourages companies to only hold what is needed for any task, which leads to very few sites holding stocks of bulk chemicals.

An understanding of the hydrogeology and local ecology of the site is vitally important. The level of the water table and any features such as rivers that may cut or be part of a site's boundaries will clearly need to be managed. Some of this is discussed more generally elsewhere in this book. All sites will require good flood planning and a strong waste water system, so that any discharge into a mains water supply is of an appropriate level. Bulk washout using water is most likely collected in intermediate bulk containers for disposal off-site. This still leaves questions such as the following:

- What occurs to water run-off?
- Is that run-off contaminated by spillage (in particular, water-soluble materials) and waste from the general surface?
- If contaminated, then at what level?
- If a flood does occur, what are the precautions available and how is flood control managed?

Clearly, questions such as those listed are only truly answerable for each specific site, but it is important to ensure that they are answered.

As well as the geographical pointers that can affect a site, there are other important considerations such as wildlife and biodiversity. Is the site unreasonably close to a site of special scientific interest? What effect will local wildlife have on the site itself? In the United Kingdom, the widespread nature and damage caused by burrowing animals, in particular rabbits, can damage both buildings and general estates facilities (e.g. piping). How can this be dealt with? Is just a rabbit-proof fence enough, or does more need doing, for example, poisoning of pests. In this case, how can this be done in a working environment without harming the workforce? Other nations have similar issues.

Waste management can be used successfully, and, in support of this, a reasonable set of models can be determined. For example, in the case of red phosphorus (RP) burning, modelling can be used to validate environmental arguments as to worst-case scenarios for the control of major accident hazards (COMAH) purposes and show how these will be mitigated for the local neighbourhood.

5.7 Production

Attitudes to the production of pyrotechnic devices are of fundamental interest when exploring the issues regarding the environment. Whilst usage will distribute the products of the combustion process (and their containers), the centralized manufacture of these items is the site where the raw materials are concentrated and the Source to pathways for pollutants that may be dispersed into the local area.

This is no more evidently so than in the case of perchlorates. Whilst the combustion of perchlorates will result in the generation of objectionable hydrochloric acid (HCl), most likely dispersed as an aerosol, the recent drive to remove perchlorates from pyrotechnics has been driven by the effects of perchlorate contamination in water supplies from nearby production sites. Any pyrotechnic item that functions without combusting most of a contained perchlorate is a poorly functioning item and unlikely to last long as a viable product. The scale of this contamination is proportional to the scale of manufacture, hence the United States is most likely to have the greatest issues with this, say compared to the United Kingdom, which produces less items using perchlorate-based materials.

In the past, the viewpoint of nearly all manufacturing industries was that when the material was purchased, all that mattered was the cost and quality of the final product. In the same manner, once a product was manufactured and sold to the customer, the manufacturer's obligations effectively ended there (except in the cases of malfunction due to manufacturing issues).

The effects of REACH, as stated, have changed this emphasis so that both supplier and manufacturer have a duty of care. This then begs the question: does the customer have a duty of care of the products that it has purchased?

The question of the environment and pyrotechnics breaks down into aspects relating to waste management and end of life (whether use in theatre or disposal).

5.8 Raw Materials Acquisition and Quality Control

As mentioned earlier, because there is such a wide range of ingredients for pyrotechnic compositions, there will also be a much broader range of specifications for use of all these materials in any one device, say, compared to that for explosives.

The first question arises with the raw materials. Assuming that a friendly REACH-compliant supplier is found, are the materials what they purport to be? If they are not, then this material becomes (and any devices made with it) simply waste. Although this sounds simple, issues arise from a variety of sources.

Modern economics and ideas of productivity has driven manufacturing along a direction of 'more for less'. While 25–30 years ago (in the United Kingdom), the checking of inwards coming raw materials and goods was controlled using receipt inspections by the manufacturers, meaning that these materials were examined and at least checked to be the specification required for subsequent manufacture before acceptance, this is currently viewed as costly and inefficient. The modern system of reliance on certificate of conformance or compliance (C of C) leads, by definition, to more waste. Although the goods-in-process is faster and 'productivity' is higher, it does not address the fundamental issue of what on earth is in the raw materials that have been purchased. Even with a reliable supplier, this can cause issues; for example, a supplier may need to find a new source of a particular ingredient, which later affects manufacturing and/or performance. An example is the changes in particulate morphology and the effect on burning rate. A metal-based colourant from one supplier may well have a different surface structure (e.g. the original source supplied a powder that is broken or fractured and a subsequent supplier provided an equally good chemical from a purity perspective, but it has rounded particles within the powder). The supplier is acting perfectly well and in good faith, but the changes in specific surface area will and do affect the burning rate or light up times for compositions using the newly sourced materials. Thus, the new material is most like to be got rid of as waste, what has been described as a 'make and be damned' approach and variable quality, leading to much more waste that requires disposal. Facilities for reprocessing such materials are generally not financially viable in mainstream manufacturing, so it is currently cheaper to simply burn unwanted raw materials. This was acceptable in the past, but it certainly is less so now.

Alongside the issue of waste, there is a further one of what are the impurities in the raw materials. This is reflected in the work of Sterba et al. [23] and the observation that arsenic- and radium-based compounds occur in civilian fireworks. Why should they not appear in those used by the defence industry?

A variant on the above-mentioned theme is that of changes in supplier's standards. An example can be given for a particular oxidizing agent. After a change of supplier, a new batch of oxidizer was shown to fail the pressure and time requirements of the customer, with the required pressure not generally being reached; and on the occasions when it did occur, it did not do so in the defined time.

Upon investigation, microscopy showed that the new material was qualitatively different. When examined by X-ray powder diffraction, the new material could easily be seen to be of much higher purity (no major impurities), whereas the original material was shown to have a number of impurities, in particular, what appeared to be remnants of starting material and also possible by-products of the manufacturing process. This simply meant that the bought material was a waste and had to be disposed of, as it would have been too difficult to attempt to replicate the original mixture. So even for an excellent supply of materials, this can also bring its own issues, given the standard recipes are dated and thus the materials they used were of questionable quality. This is an interesting inversion of the impurities problem.

While these examples highlight materials that may be synthesized in simple chemical reactions, the quality (and thus waste) issues go across the board, there are also problems with the usage of natural ingredients, particularly in coatings and binders.

For the majority of the history of pyrotechnic usage, many of the organic ingredients were derived as natural products, e.g. beeswax and acaroid resin. However, the difficulties in maintaining any form of quality control and continued performance led to a drive for new synthetic polymeric products being used, e.g. resins and varnishes, to reduce the variability in materials performance that often occurs with natural products. However, although quality and product control is improved, the nature of and the ingredients used within those modern synthetic materials means manufacturing has a different set of environmental issues to address, the products of combustion can be problematic, and the resultant materials may be of environmental concern. In particular, this arises for polymeric binders which use halogens within their structure.

These are typical of the issues around acquisition of raw materials and the questions continue regarding the purity and quality of the raw ingredients, in the case of the widely used amorphous boron:

- What is the quality control of boron?
- What are the real-size requirements of the amorphous boron?
- What percentage amorphous need it be?
- Is this information accessible by the currently used simple tests?

In the case of sulfur for pyrotechnic items, the purity can be as low as 98%. If it works at this level of purity, then from a cost perspective it needs to be used; but from a quality (and thus waste perspective), is it good enough? It might be feasible to go as high as 99%.

One issue that the pyrotechnics industry has to address in this respect is its attitude to control of supply. As a low-volume customer of their suppliers, there is pressure as manufacturing supply is responsive only to major users, e.g. the plastics industry; nevertheless, it can do something regarding the quality of a number of materials. Reprocessing adds to the cost of manufacture; but if it is incorporated from the start, then savings are likely to be made in the longer term. The pharmaceutical industry also claims to be a low-volume user for many similar (and some identical) chemicals, but it knows it has to have stringent quality checks and reprocesses many of the inward coming raw materials.

5.9 Specific Materials Production

During manufacture and usage, we must examine the possible pathways by which a pollutant can move from source to receptor. These are most likely to be the dispersal of solid particulates and liquids, the interaction of solids and liquids with water, and the dispersal of gaseous or airborne products of combustion. What can be most problematic are those materials which are soluble in water or are volatile, which will have a greater chance of being carried far from the source. This also means that their dilution will be greater, but given that even small amounts of pollutant can have major effects on ecology, these data are important to consider.

5.10 Heavy Metals

The focus on heavy metal contamination at production sites concentrates generally on the elements lead and chromium. Hexavalent chromium compounds are already on the REACH authorization list, although it is of interest that the only current lead compounds are those associated with chromium. However, an examination of candidate substances for future addition to the authorization list shows a broader range of lead compounds. Although in the United States there is movement to replace barium within the EU, it is clearly seen as less of a problem and strontium is seen in a similar vein. The use of lead and chromium is at relatively low levels in any one device, most commonly found in initiation/igniter systems and delays. However, barium and strontium are used more widely and in larger volume as part of coloured flare systems. There are still legacy igniter systems being used which contain such unpleasant compounds as lead chromate, but one can only assume that these reside within legacy items that will either be replaced soon or at least hopefully will not be used and disposal required at the end of service (see later). Lead and chromium are still found in delay compositions, although air countermeasures have already moved away from them. However, lead is still found in percussion caps. The old red lead/silicon primer is also still used – it is inexpensive and seemingly very effective, but it is on the candidates list and permanganates are also still used.

From a solubility perspective (and thus ease of transportation by water), heavy-metal-containing salts do not show any real trend in increased solubility. However, it is at least noticeable that the majority of chromium salts are generally of low solubility, the most soluble being sodium, potassium, magnesium, and ammonium chromate, but the dichromate salts (as coatings) are more soluble.

As well as the compositions themselves, it should not be forgotten that these materials occur in pyrotechnic devices as parts of the surrounding engineering, with chromium acting as passivation in cases (Al/Cr cases) and the dated, but still used, swaged lead delays. It is thought that this may have more of an impact than changing compositions and, of course, as soon as a change to even the engineering is made, commercial customers will need to be informed and appropriate tests performed. This is likely to occur on a relatively large scale, so at some point, the customers will have to take some notice. Other metals do appear as pollutants, and in fact in the case of zirconium, used as part of armour piercing and incendiary rounds, that this was discovered as a pollutant was due to monitoring of the water supply. Indeed, most environmental issues were discovered due to being picked up in the water supply.

In parallel to heavy-metal-based pyrotechnics, they are closely aligned at the igniter stage with primary explosives, which although not the topic of this chapter, should be mentioned as a major waste producer. As a process, the manufacture of lead-based primaries is inefficient and wasteful, but given that these are candidate materials within the REACH SVHC list, this may prove to be a blessing.

5.11 Perchlorates and Chlorates

As opposed to the United States, there is much less concern regarding the manufacture of pyrotechnic systems using perchlorates in Europe. Indeed, there are no perchlorate or chlorate items in the authorization or candidate list. This most likely reflects the lack of bulk manufacture of the materials in the EU, rather than in the United States. The major issues of the chlorates and perchlorates are both the amounts that are used on specific sites and also that they are relatively soluble, leading to leaching into the local water supply.

5.12 Smokes

In the area of smokes, given that the toxic obscurant hexachloroethane (HCE) smoke systems are now obsolescent and are no longer manufactured, there is little to affect what should be good manufacturing practice. The most pressing concerns are the quality of the RP used in obscurants (e.g. what percentage of it is white phosphorus and will it affect manufacturing and that of supply). A second concern is the ability of manufacturers to utilize solvents in the downstream processing of the associated rubbers. Again, a number of organic solvents have been removed from use due to REACH; and although there are replacements, they will change the processing of the associated product and are themselves possible candidates for future REACH action. The third main concern is that, as mentioned previously, the lead in RP manufacture (and dyestuffs) has come out of the plastics industry. Thus, the biggest risk in the waste management chain is the continuation of source and maintaining the same source.

5.13 Volatilization Smokes

It is now well known that the dyestuffs associated with volatilization smokes are the most pressing concern for the pyrotechnics industry. Additive materials such as cinnamic and terephthalic acids and the variety of coloured dyes are not of great concern at the manufacturing point of view – these need to be controlled as any powder manufacturing needs to be. However, the usage does require further examination and will be done so later in this chapter. The manufacture of CS does require more concentrated efforts, but this can still be done in a reasonably efficient manner, even local protection is required.

5.14 Magnesium Teflon Viton (MTV) Countermeasures

In the processing of magnesium Teflon viton (MTV), there are a number of areas which can cause environmental issues. A number of these have been placed on the list of authorized materials and it would be no surprise if more were to be so. In addition, the use of organic solvents such as butanone, hexane, cyclohexane, and acetone may cause local effects – it is expected that hexane will be completely replaced by cyclohexane to minimize toxicity and mutagenic problems. Nevertheless, there are problems with local exhaust ventilation (LEV) systems. In particular, where there are no scrubber systems in place, as this causes some of the organic vapours to be dispersed into the local environment. Although it would be preferable for LEVs to employ a scrubber system, it might yet be too expensive to either replace the existing ones or to build a new processing plant containing them.

5.15 Resins, Binders, and Solvents

As mentioned earlier, there is a strong move to remove isocyanates from the manufacturing cycle. In fact, this is probably the most pressing example of where movement is occurring within the European industry. Whilst the types of isocyanates used nowadays are less volatile than those of earlier times, they still can affect the health of those around them. Even the seemingly benign polytetrafluoroethylene (PTFE) has been shown to have health effects. A range of different types of process are being investigated such as less toxic curing agents and thermosetting plastics. Spillages of liquids are still a problem in production plants, given that in addition to these organic chemicals, others such as dichloromethane (DCM) and amyl and butyl acetate are used. Mistakes are still made which can have long-lasting effects. An example being the old Wallop Defence Systems site in North Hampshire, where a number of years ago TCE was used to wash equipment and there was a leakage into the surrounding area. The movement of the TCE is still being followed in the local geology. While it shows no real effect on the local water system, the added expense of boreholes and continued monitoring is always unwelcome.

5.16 Storage

We would like to think that during the short period of storage within a manufacturing plant (prior to delivery to a customer) and whilst the customer stores the product prior to its use, that, as long as storage conditions are adequate and monitoring continues, nothing will happen. However, there are certainly situations where even in the best monitored situations degradation of a product will give rise to environmental issues. These are, in general, associated with manufacturing issues, a case in point being the RP smokes. If there are issues in manufacturing, then it is possible that the munition containing the RP will start to generate phosphine, which is reactive, toxic, and explosive. It can eat through the casings (e.g. eating away at the weak points such as fuzeheads [35]) and can be released into the locale where people are working [36]. While phosphine is reactive enough to disperse and break down quickly in the environment, it is still relatively commonly used as a pesticide and fumigant under strictly controlled conditions. Thus, it will affect the local area significantly. Most munitions are likely to have a lifetime of over 10 years before going to disposal, so the concept of slow degradation does need to be kept in mind as part of a clear Environmental Impact Assessment on a MoD store, for example. Other nations have similar issues.

One way forward that has occurred recently has been the study by Koch and Cudzilo [37], where they examine the role that phosphorus(V) nitride (P_3N_5) can play as an ingredient for an obscurant to replace both RP and white phosphorus.

P_3N_5 is a stable compound, and shows no reaction to the usual stimuli associated with pyrotechnic compounds. Unsurprisingly, P_3N_5 does not undergo any real decomposition until at least 700 °C, an endothermic process, mimicking the preferred use of phosphorus-based compounds as anti-fire agents in the plastics industry.

During decomposition, P_3N_5 undergoes the following reaction

$$P_3N_5 + \text{heat} \xrightarrow{\text{yields}} P_2(g) + PN(g) + 2N_2$$

liberating phosphorus gas, which readily reacts to form the obscurant phosphorus oxides. Investigations into formulating P_3N_5 with green oxidizing agents have led to compositions which have a yield factor of about half that of RP, but with a much higher figure of merit. In addition, the authors state that P_3N_5 does not evolve any phosphine when directly compared to microencapsulated RP.

5.17 Packaging Waste

The packaging of stores can also be questioned. Is a store correctly packed to achieve an appropriate travel certification, or is it simply done in a manner that is easy and requires no thinking? How much simple waste could be removed by correct design and manufacture? Whilst packaging is part of a munition's design to effect safe transportation and storage, it may well be that extraneous packaging is being used simply because it can be. To some degree, the thinking behind this is that if a particular packaging works for an item, then doubly (or triply) bagging the item will enhance the safety to a point where it is completely safe. The reality is that they essential packaging will act as mitigation for any safety concerns and adding further packaging is simply wasteful.

5.18 Usage and Disposal

We may consider the usage and disposal in a manner similar to that in the Manufacturing section of this chapter.

5.19 Heavy Metals

In use, the heavy metal products of theatre use are most likely to have a relatively minor effect. This is due to the fact that, presuming the munition functions as intended, the heavy metal will have converted into whatever the product of combustion is and also that it is more than likely to be spread out over a larger and moving area. This is not intended as an apology for not moving to more benign materials, but a description of what happens. In addition, for the most objectionable materials such as Pb and Cr, they are used in small amounts within any device, so the mix of small amounts and a spread of the materials in a moving theatre will minimize the environmental impact.

However, where this does impact on the environment is where there is no relative movement of the site of exposure. In particular, testing areas, training ranges, and disposal grounds may be more directly affected by the long-term continued accumulation of small amounts of heavy metal by-products.

Ranges are discussed much more fully elsewhere in this book, although not specifically with regard to pyrotechnics. However, a number of points can be raised here. Firstly, they are in continual use by the energetics community, so any dispersed material will be concentrated at these sites. How the material is then dispersed depends on the geography of the site, for example, what the water course wash-off is like, what the local weather system is like, etc. And it is important to remember that not only will there be combustion products but there will also be unexpired and misfired items, which will be degrading and ultimately leaking their components into the environment.

In general, although the above-mentioned is true, there are some legacy items which may well concern even those using munitions in a moving theatre and that is the fact that there are still some legacy devices which contain the highly toxic thorium and thallium.

Whilst this is effectively being ignored by the UK/EU axis, continued support by the US DoD has meant that there have been some steps forward since the reviews of Sabatini.

The work sponsored by the US Army at Armament Research, Development and Engineering Center (ARDEC) has examined two particular routes for the development of new delay compositions; either thermite formulations or the use of B4C within delay compositions.

In the case of the thermitic approach, this has taken the route of using MnO_2 as the oxidizing agent. One approach uses manganese itself as the fuel, the other uses tungsten.

For the work by Miklaszewski et al. [38] on Mn/MnO_2 thermites, it is the richness of the Mn_aO_b system that allows this area to be used. The estimated reaction appears to be

$$x\text{Mn} + y\text{MnO}_2 \xrightarrow{\text{yields}} z\text{MnO} + \left(\frac{1}{2}y - \frac{1}{4}z\right)\text{Mn}_3\text{O}_4 + \left(x - \frac{1}{2}y - \frac{1}{4}z\right)\text{Mn}$$

This broad study showed that the temperature achieved by the Mn/MnO_2 system ranges between 1400 and 2000 K, depending on the stoichiometry of the sample and that initial safety studies indicate that this composition has excellent electrostatic and friction sensitivity values, as well as producing only a small amount of gas, compared to traditional delay compositions. As if this was not good enough, the same group [39] moved to examine the W/MnO_2 system. In this case, the suite of reactions predicted is as follows:

$$W + 3MnO_2 \xrightarrow{yields} WO_3 + 3MnO$$

$$WO_3 + MnO \xrightarrow{yields} MnWO_4$$

$$W + 3MnO_2 \xrightarrow{yields} MnWO_4 + 2MnO$$

However, given this is one of the better studies of the combustion products, it appears that the final reaction given is not the complete picture. Two further routes could be occurring:

$$3W + 7MnO_2 \xrightarrow{yields} W + MnWO_4 + Mn_3WO_6 + Mn_3O_4$$

$$2W + 7MnO_2 \xrightarrow{yields} MnWO_4 + Mn_3WO_6 + Mn_3O_4$$

However, neither of these two reactions can define the final composition. The full story is likely to be a combination of these (or related) reactions, combined with a more accurate estimation of the combustion products, although it is based on the relative levels of stability of the various oxides of manganese. The depth of study is not quite as full as that for the Mn/MnO_2 composition. While the W/MnO_2 study does already move to examining the formulation in the context of a delay, there is no sensitivity data given. However, this also appears to be a good possibility for another green gasless delay composition.

The second route that has been taken, and seemingly successfully so, is that of the versatile B_4C. An initial study in handheld signals showed that for their current design, the only delay that can be used is a $W/Sb_2O_3/KIO_4$, as the others extinguish, rather than continuing to combust. However, while this composition has the advantage of replacing classic perchlorates with a periodate, the composition still contains antimony, which is clearly a toxic heavy metal and pollutant [40].

In response to this, the work of Shaw et al. [41] showed that $B_4C/NaIO_4/$ PTFE will act as an excellent candidate to replace the original composition. As described, both B_4C and $NaIO_4$ are considered 'green'; and although PTFE has a long history of poor manufacturing control, recent developments in the fluoropolymer industry have made this a more favourable energetic binder to use. The work has proceeded to the point where it has moved from a study piece, to being demonstrated to work in the US Army handheld signal munition (Table 5.2).

The final change in this area is the development of calcium sulfate as an oxidizing agent for Si fuels in delays. This work of Tichapondwa et al. [42], developed by the South African Mining Industry, is driven by civilian uses, rather than a military context. However, the outcome is the same. Calcium sulfate is a well-known

Table 5.2 A comparison of dynamic behaviour of B_4C-based flares to W-based HHS flares in conditioning trials.

Conditioning (temperature)	B_4C delay dynamic time (s)	HHS delay dynamic time (s)
Hot (71 °C)	4.77 (0.29)	4.87 (0.22)
Ambient (18–22 °C)	5.83 (0.42)	5.20 (0.15)
Cold (−54 °C)	7.02 (0.39)	5.78 (0.17)

Source: After Shaw et al. 2015 [41].

oxidizing agent, but it has only recently been thought of as a 'green' oxidizer. Issues still exist with the use of $CaSO_4$, as the material decomposes via the following reactions:

$$CaSO_4 + 2Si \xrightarrow{\text{yields}} CaS + 2SiO_2$$

$$CaSO_4 \xrightarrow{\text{yields}} CaO + SO_2 + \frac{1}{2}O_2$$

$$Si + O_2 \xrightarrow{\text{yields}} SiO_2$$

$$SiO_2 + CaO \xrightarrow{\text{yields}} CaSiO_3$$

This indicates that the major gaseous elements are likely to be sulfites or possibly sulfur trioxide, all of which are unpleasant. However, the predicted software indicates that the main gaseous products are most likely S_2, SO_2, and SiS.

The continued new avenue of synthesizing metal salts for use as colourants continues. The Klapötke group is now well-known for systematic synthesis of metal salts. In addition to metal salts of 5,5′-bis(1-hydroxytetrazole) [43] and 4,5-dinitro-1,3-imidazole [44] for use as colourants, replacing heavy metal species, more recently, they have developed the lithium salt of 3,3′-diamino-4,4′-dinitramino-5,5′-bi-1,2,4-triazolate [45]. In this particular case, the high nitrogen content minimizes the oxidation of the lithium atom, permitting it to remain in the gaseous atomic state for long enough to be a practical red-light-emitting source.

Along a similar vein, the Klapötke group has also pursued the high nitrogen route in their search for new infrared illuminating compounds, utilizing potassium and caesium organic salts [46]. However, these studies are like many high-nitrogen-based systems, still at the low technology readiness level (TRL) stage, and time will only tell if they move to being more developed and move to use.

At a relatively advanced stage is the replacement of the R440 dim illuminant by a new composition, termed R680 by Moretti et al. [47]. The R440 contains barium and the new formulation replaces this with strontium. The tracer appears to be functioning appropriately, but further tests (e.g. spin tests) are needed to see if this composition will be a direct map to the old R440 composition.

5.20 Perchlorates and Chlorates

As mentioned previously, it is generally the production of these materials which poses environmental questions. However, again, in the past few years, there have been steps forward in the performance of perchlorate free materials. The most forward movement has come from the work of Miklaszewski et al. [48] at the Naval Surface Warfare Center, Crane, in moving a novel yellow-light-emitting composition to a useable item, in particular the marine Mk144 illumination signal. This work uses established pyrotechnic composition analysis to remove the objectionable $KClO_4$ and replace it with the benign $NaNO_3$. A set of 13 compositions were studied in an effort to change the mechanism of yellow light generation. They accurately described the common emission bands associated with the product species generated by the Mk144 (Table 5.3).

Of the 13 compositions analysed, three proved to have suitable burn and optical parameters for usage within the Mk144 flare and are listed subsequently (Table 5.4).

A bonus from an environmental perspective is the discovery that one of the flare compositions (9B), in addition to removing the perchlorate, also manages to replace the barium that exists within the current Mk144 composition. This means that the light generated is simply that associated with the gaseous Na from the oxidizer and flame expander, and that is only residual green light from the MgO blackbody radiation. For this particular composition, the apparent burn time, average luminous intensity, and luminous efficiency all exceed the requirements used in the current composition and the ignition safety characteristics (impact, friction electrostatic sensitivities) are all comparable to the current formulation. With probably further tests to come, this may yet prove to be an excellent replacement and it is encouraging that this still can be done within the context of classical pyrotechnics.

Table 5.3 Common yellow-light-producing flare emission bands, excluding blackbody contributions.

Species	Emission wavelengths (nm)	Visible colour
BaCl (g)	512.9, 524.1, 532.1	Green
BaO (g)	549.3, 564.4, 586.5, 604.0, 649.4	Green, yellow, red
BaOH (g)	487.0, 512.0	Green
MgO (s)	498.6, 499.7, 500.7	Green
Na (g)	589	Yellow-orange

Source: After Miklaszewski et al. 2017 [48].

Table 5.4 Successful yellow flare compositions from the Naval Surface Warfare Center.

	5A	6B	9B
Mg (Gr 17)	23.00	19.67	25.5
$Mg_{0.5}Al_{0.5}$	3.60		
$NaNO_3$	27.40	36.20	38.2
$Ba(NO_3)_2$	29.0	26.47	
$Na_2C_2O_4$			29.4
PVC	8.10	10.66	
Asphaltum	3.95		
Epoxy	4.95	7.00	7.0

Source: After Miklaszewski et al. 2017 [48].

In a similar vein, the simple compositional rearrangement approach has been utilized in a theoretical paper from Brusnahan et al. [49], where they critically examine the potential replacement of perchlorates with periodates in a 'drop-in' manner, where $KClO_4$ and $NaClO_4$ are replaced by the periodate analogues, KIO_4 and $NaIO_4$. In this theoretical paper, they examine the physico-chemical properties of the periodate analogues and associated decomposition pathways.

Periodates appear to have a number of beneficial properties, such as a more negative enthalpy of formation (and thus greater stability) than their perchlorate cousins. They also have advantages in that there is reduced hygroscopicity of the examined periodate species. However, there are clear issues with a direct replacement. Periodates seem to have chemical compatibility issues compared to perchlorates with two commonly occurring materials, tungsten metal and copper salts. In general, chemical compatibility is a measure of the stability of chemical mixtures. For example, in a pyrotechnic composition of a fuel and oxidizer, if a replacement oxidizer is required, then it must be ascertained that the new oxidizer neither changes the reaction rate (whether slowing it down or accelerating it) nor should it spontaneously react with the fuel. In the case of $NaIO_4$, this can cause an oxidation with Cu(I) in the novel lead-free primary material, DBX-1, leading to the creation of $Cu(IO_3)_2$. Tungsten reacts slowly with KIO_4 to leave a brown discolouration (I_2). In addition, the decomposition reaction of the periodate materials is ultimately endothermic, as opposed to the exothermic decomposition of the perchlorates. This requires careful balancing and design for any pyrotechnic system, as does the appropriate mass balance through use of heavier periodates. However, there are still some concerns regarding the toxicity of periodates (which is currently under study), but there is hope that a simple 'drop-in' replacement can occur, at least for flash compositions.

While much effort has been expended on the removal of perchlorate from pyrotechnic systems, there have also been moves to further reduce the pollutants derived from pyrotechnic use. Many products of pyrolysis can be either toxic or carcinogenic, and it has been understood for many years that unpleasant materials such as polychlorinated biphenyls can result from pyrotechnic devices containing chlorine. As such, a number of studies are now moving to examining the removal of chlorine from all pyrotechnic systems.

In most cases of perchlorate or chlorine removal, it has been primarily the oxidizer species which has been studied – not unreasonably as historically, colour has been introduced to the system either by the oxidizing agent, or by a separate colourant. However, Brusnahan et al. [50] have shown MgB_2 to be a possibly pyrotechnic fuel, which has the propensity to emit green light on combustion of appropriate spectral purity and is seemingly insensitive to different ignition stimuli. However, there are still issues with this material in possible use as a fuel, burn rate control and binder addition being two simple examples. While boron substitutes have even moved into the realms of direct use, a number of further studies have examined boron esters and boron salts as a source of the metal [51, 52].

5.21 Smokes

5.21.1 Obscurant Smokes

The removal of HCE smokes from service has ameliorated the worst effects that they have on health and the environment; and RP is now the obscurant of choice although unsuccessful efforts have been made to find alternatives. It needs to be remembered that it is not true that it is non-toxic. It is of lower toxicity compared to HCE and therefore still requires thought as to how it can be used in a manner that will least affect the environment.

5.21.2 Volatilization Smokes

Coloured and signalling dyes. Within the United Kingdom, as with the United States, the most pressing concern is that of the toxicity of the coloured dyes used within a signally smoke. A number of them are toxic, and work is ongoing to examine the toxicity of the others. In addition to the issue regarding toxicity, the other major problem is that of respirable particulates in the immediate environment. The efficacy of a smoke is dependent on a number of factors, and a major one is that of particle size – for nearly all smokes, the optimum particle size is the same as that which can be inhaled easily into the lungs and remain there. There have been studies to look for alternatives, for example, an industry-funded academic study [53]; but at the moment, no real alternatives have been found to the dyes currently used. From a broader environmental perspective, it may be negligible; but it may well be a health risk for repeated users as the full breakdown chains for the dyes are as yet unknown. Unfortunately, as the dyes are generally organic in nature, it may yet well be the case that the more science and testing is done, the more issues are found with each material used as a dye.

The difficult situation is the testing of such devices; smoke products are generally used in the open air and this is how they are largely proof tested; therefore, there may be increased concentrations of the combustions products around the proofing grounds. In some cases, due to the nature of the smoke, this may involve the use of test ranges away from production facilities and this leads to reliance on available ranges and their pollution restrictions.

However, the UK suppliers have worked on REACH issues with the dyes that are currently used and registered in REACH – and there is at present nothing of high concern.

Nevertheless, there are a number of areas where advances have been made in the past few years. One aspect of the production of handheld coloured smokes (for signalling) is the requirement for the use of organic solvents during manufacture. The removal of these solvents would both render the pyrotechnic more economical to make but also cut through the environmental footprint for that particular item. Further work at the US Army at ARDEC has shown that such solvent free manufacturing is possible. The group of Moretti et al. [54, 55] managed to develop both a prototype and demonstration system for a solvent-free yellow smoke, utilizing solvent Yellow 33 and comparing it to the current in-service Mk194 smoke and a battlefield simulator.

The approach of Moretti et al. is an excellent demonstration of how to approach a single task and create solutions that support a broader range of issues.

As well as examining solvent-less smoke composition creation, the same group has examined the case of white smoke [56]. The loss of HCE-based smokes from the US Army's inventory has caused problems. Although replacements existed, in the guise of both cinnamic and terephthalic acid–based white smokes, these latter two were originally only developed for signalling and training and did not meet the requirements of the Army.

The group at ARDEC utilized B_4C to create a suitable white smoke generator [57], with associated environmentally friendly components. Typical compositions of B_4C/KNO_3/KCl/Ca stearate/PVAc were studied and an excellent smoke generator was developed. In particular, the yield factor for these B_4C-based smokes was nearly double that of the CA/TA-based white smokes, while not quite as efficient as the older HC smokes.

Alongside the work carried out by the ARDEC group, Wilharm et al. at NSWC Crane has performed a theoretical study of the compound K_2FeO_4 as a possible green oxidizer [58]. The study utilized the NASA-CEA code to identify if K_2FeO_4 could sustain combustion; and, indeed, it could. In fact, it appears to be a strong oxidizer and implies that care must be taken in making sure that it does not degrade any materials with which it is in contact, although further work needs to be performed.

5.21.3 MTV

In theatre or air tests, the flare compositions are released over a wide geographical area, both from a perspective of speed of travel of the dispensing aircraft and the altitudinal variations, compared to the much more geographically static usage in sea or land theatre. Koch has described the manufacture of toxic products of combustion of MTV [59] and points out that the most toxic possible by-products (e.g. HF and AlF_3) are calculated to be released at low concentrations and are extremely likely to be dispersed into even lower concentrations due to the speed and trajectory of the aircraft and dispensing system. In particular, it ought be noted that HF is lighter than air (compared to, say, CO of HCl) and is less likely to 'settle'. However, as with most other compositions, it is at localized usage sites that the worst pollution occurs, in particular at testing grounds where trials are conducted.

5.22 Disposal and Waste Burning

Where feasible, disposal ought to be performed by an accredited method. External contractors can be found who will remove waste materials such as waste, and residue CS can be removed and destroyed by explosion if it is dissolved in water. In a similar manner, gelatinized nitrocellulose can also be removed and disposed of by contractors, as are any completed stores. Of course, this relies on the contracting company behaving in a suitably environmental manner.

Although this is a method for disposal, the majority of waste is disposed of by local burning at the site of manufacture and is by far the biggest concern for current manufacturers. While requiring an expensive licence to be permitted to burn waste products, it is currently the only real method for waste disposal for many items, whether compositions, failed devices, or contaminated waste. Although there are as yet no plans for a unilateral ban on burning waste, the likelihood is that it will become increasingly restricted and the problems that this could bring means that the focus of industry is very much on alternatives.

Minimizing any burning on-site is a vital aspect of this process and this is part of a broader waste management system. The maintenance of a proper waste disposal log and monitoring of the disposed items will mean that environmental impact is as low as can be. For instance, low-volume waste burning is based around compositions and items, whilst high-volume waste tends to be those materials which are contaminated, e.g. personal protective equipment (PPE), swarf, and discarded components.

Restrictions already apply to regulated burning, for example, because of wind and weather conditions, as well as the local surroundings (as mentioned earlier). COMAH restriction means that it is vital that smoke clouds are kept to a minimum. Again, for example, with the United Kingdom or The Netherlands being of small size and heavily urbanized, this means that major traffic arteries can be close by. There have already been accidents on motorways due to unregulated agricultural burning and clearly no one wishes to be the cause of another similar accident.

One method to ameliorate this is to keep burn quantities to as small as possible and move to smaller regular burns for waste disposal. During the burn, if it is possible to get the temperature of the burn high, then it will minimize the burn time, getting it over and done with quickly and also means that the maximum amount of solid waste will be consumed. In general, CO_2 emissions are low, but can be reduced by cutting down the amount of fuel oil or diesel used. There is a move away from using diesel for burning, due to its black smoke and particulate production, with many countries attempting to reduce diesel emissions, whether from vehicles or other sources. To replace diesel, there is a move to using triacetin for a cleaner burn. This has the added advantage of having a higher flash point;

and thus when it burns, it does so much more completely. It is also seemingly more environmentally benign and the material itself is very biodegradable.

In the specific case of burning MTV waste, this has already been described by Koch [59]. Whilst the work is an excellent basis for the study of burning, there are a number of factors which have been overlooked. There are some theoretical studies of the burning of MTV burnt with diesel and these have shown that MTV will generate hydrogen fluoride and metal fluorides, magnesium fluoride and carbon soot as part of smoke formation. Koch states the important fact that in the open air, the dispersal of these products is much greater than in an enclosed environment, such as testing tunnels, so that monitoring is very important in these enclosed spaces. But there are issues with this. As mentioned earlier, it is generally not just MTV and diesel that is burnt. The discarded PPE, swarf, and other contaminated items consist of large-volume burning materials and these are not easily modelled, but will contribute to the local impact environmentally. As Koch states, there has been one study of burning waste [60]; but given the changes in both understanding and technology in the 20 years since that study was made, more is urgently required.

In addition to MTV, the burning of waste chlorinated materials could generate HCl and other unpleasant organochlorine pollutants. Evidence for and against this is needed with determination of quantities and the hazards. In the case of DCM, this is commonly removed in cloth bags and burnt. But it is done so in a clay-lined burning ground, which acts an impermeable lining and stops leaching into the soil.

In the past, there was almost uncontrolled burning of waste, enhanced by the use of diesel, which was and still is commonly used as a desensitizer for waste pyrotechnic compositions. Although the burning is now much more strictly controlled, it can still have problematic effects. Care also has to be taken, in that like football players and diving, waste burning is 'theatrical', to show that hazardous waste is being got rid of by the companies. But this theatricality must not be at the expense of the workforce or the local environment. There has to be a reason for burning and a duty of care to everyone during such times.

5.23 The Future?

The ongoing pressure of continued environmental legislation and regulation means that whatever is occurring now will need to change. Although the rate of change is low, so is any impetus to change; in addition, the defence sector sometimes does not understand the implications of legislation, during the original moves to REACH implementation, and the knock-on effect for suppliers and manufacturers were not necessarily understood by the customers.

While it is possible to simply substitute one chemical for another, this can still create problems, whether it is the cost of implementation within a manufacturing process and subsequent requalification of a selected store, or the actual physico-chemical effects of the substitution. For example, one particular tracer composition originally used TCE as part of the manufacturing process. As REACH precluded the further use of TCE a simple substitution was made, bringing in DCM as a replacement for TCE. However, this simple replacement has had safety and failure implications. The DCM evaporates more quickly than does TCE, which although speeding up the manufacturing process by allowing the pyrotechnic component to dry more quickly, also has the effect of leaving it much more brittle and more granular than the composition manufactured using TCE. This granular composition is then much more sensitive to accidental ignition. By banning TCE outright, no questions were ever asked of how much of it was used, where it was used, when it was used, or if it could be recycled. Given that there is an expectation within the industry that DCM is also likely to be banned, it merely causes a new generation of headaches.

The scenario described may sound as a plea to roll back REACH and reintroduce TCE for manufacturing processes, but this is not the case. The plea is for better investment and research into the processes and materials that we use, so that truly suitable substitutes can be found, not just convenient ones.

This may seem a simplistic scenario, but the increased failure rate of the composition and accidental ignitions (leading to devices that require disposing of) leads to a greater amount of waste (which itself is hazardous and environmentally questionable), which then does need disposal. That disposal, which is most likely to be open burning, will then produce noisome smoke and gases, as well as ash that may well have residual contamination, bringing its own set of questions:

- Is any residual ash contaminated?
- Are all materials in the process legal?
- What are the environmental legal concerns about an accidental ignition?
- How have the changes in acquisition of raw materials changed?
- How has manufacturing changed?
- How has usage changed?

The manufacture of pyrotechnics and their design is strongly dependent on the understanding and experience of a number of specialists. However, their number is rapidly dwindling.

5.24 Suitably Qualified and Experienced Person (SQEP) Issues

Within the United Kingdom and across the EU, it is recognized that the defence industries are losing expertise and knowledge, both at the commercial and at the institutional level [61]. The number of personnel with appropriate experience and understanding of the creation of energetic systems in general and more particularly, pyrotechnic systems is rapidly decreasing. While the reasons for these losses are varied, if care is not taken, this will lead to a major gulf not only in general manufacturing but also in specialized areas, such as pyrotechnics. Without the specialists in both academia and industry, even if there is an impetus to provide and test new compositions and develop new ideas, there will be no one with the appropriate expertise. This will require nations and companies to invest in the future training and development of personnel, as well as making the sector more attractive as employers. The issue of suitably qualified and experienced person (SQEP) is affecting the whole of the energetics industry, but it is much more keenly felt within the pyrotechnics community, in that the range of materials and required effects in the pyrotechnics sector is so much broader than the explosives or propellants sector, that there is simply much more to learn and understand.

Alongside the SQEP issues, the effects of economic change and models of production have led to loss of personnel. In earlier days, a working team would consist of a workforce of ~15–17, alongside six quality control staff; this has now reduced to ~9–11 members of staff and no dedicated quality staff. Where are the eyes of control and quality? As described earlier, it is currently cheaper to 'make and be damned' and in addition, the control of quality cannot be a high priority for the customer. Thus, an increase in waste and thereby in pollution is driven by 'economics' and a lack of importance to either the manufacturer or customer. All in the name of 'economy'. This may only change when it becomes 'uneconomic' to simply dispose of, or burn waste.

Of course, one answer is to automate a system; however, this always requires a very large investment in capital equipment, retraining of staff, redefinition of materials quality (both incoming and during processing), and requalification of the entire catalogue of products to be made. In addition, the relative regularity of ignition events during manufacture means that there are always going to be times when either the entire plant or parts of it, will be 'down' while investigations take place.

There are moves to automate much manufacture, but it has proved to be expensive and this ought to be part of the answer. Another aspect that can be followed is to examine and correspond with other sectors that manufacture systems that may work in a similar manner; for example, there are many similarities of different parts of the pyrotechnic production process to those in the pharmaceutical, foodstuffs, and petrochemical industry. We should talk to these and others that can offer insights.

5.25 Integration

What is required is an integrated and holistic approach to pyrotechnics. In effect, a systems approach to the manufacture of pyrotechnics and devices, not just from a construction point of view but from a whole item and lifespan perspective. Where are the materials coming from, what is in the materials, how are they processed (if at all), what waste is produced and how is it removed, how are items transported and then stored and what happens at the end of the munitions' life? Whether this is the usage, or the breakdown and destruction of unused stores?

The effects of all environmental legislation (whether REACH or other similar regulations) means that all sectors are being forced to re-examine their processes and all are being forced to examine the whole life cycle. This does not mean that all aspects need to be assessed under the same regimes, and this systems approach should blend together the different aspects of the whole life of a device. For example, research is very much of the moment and uses small quantities, whereas end of life is not really thought about.

Different aspects of the production process can be looked at with new eyes; in chasing performance, you want to use the plateau of the burning rate, rather than the peak of the burning rate, so a simple substitution for a less energetic mix might be suitable. With regard to toxicity, what can you change? And can you make an assessment at the beginning of the project as to whether there is a replacement available? How much of the process or the materials can simply be disposed of?

To manufacture pyrotechnic devices or items in an economic and environmentally friendly manner, there are a number of things that need addressing, some of which can be done so by any one individual company or organization, the other is a longer term approach.

From a corporate perspective, a systems engineering approach, experienced personnel, and a good upfront design throughout the life of the device are vital to be able to develop and then manufacture a pyrotechnic item which will behave in an environmentally benign manner. The second issue that for a forward-looking and coherent development approach, the defence community has to act in a more cohesive manner, which has been forced to be diverse due to political/commercial pressures, never mind the issues regarding the loss of SQEP expertise.

There is a duty now to invest for the future, and industry needs to be much smarter at the research and design stage. Not just to bring in the brightest minds to develop new products but to do this with other bright minds, so that an integrated whole comes from the diverse parts. Pyrotechnics is unusual in that it has always required an understanding of both the chemistry that is occurring and the engineering in which the reacting chemicals are housed. Engineers and chemists need to work closer together and understand each other and to design better items. One has to consider the whole device, from manufacturing, through storage and use (or disposal).

- What is it that you want to make?
- What does it do?
- How will you make it?
- What are the environmental properties of the raw materials, intermediates, final products, and post-use residues?
- What happens to people in manufacturing?
- What happens during storage?
- What happens during use?
- What are the traces in the environment when it has done its job?
- Why does industry not make only what is required?

These points will require a much more integrated approach from both customers and industry. The manufacturers will need to make devices in an environmentally benign manner and it is equally important that the customers need to be considerate to producers, to understand what the components of a store are and what the whole picture of manufacturing is. This broad and integrated approach requires open minds and for all parties to gather and discuss the issues at the start of any project. The customer may well be initially concerned with the environmental case of firing a store, but in requiring that, they need to understand and appreciate what has to go into that requirement. This raises the question of whether the customer has a duty of care for the manufacturer to make products in an environmentally safe manner. Otherwise, why not simply buy a product from the cheapest source?

The whole sector has not helped itself by being very conservative. While there are excellent reasons for this, the changes in both science and regulation means that change will come. So, for example, it is cited that the conservative outlook stems from a concern about materials sensitivity and the ability to compare one material against those already known. This is perfectly fine, but that must have started somewhere, so why not start a similar environmental register?

These channels of communication are vital and are often the weakest link in the chain. There has to be a frank and open exchange of ideas between both manufacturer and customer. An old-fashioned attitude of 'I do not care how you make it as long as it works' does not hold water any more.

There are some very interesting moves to replacing current materials with novel and less polluting ones, e.g. the replacement of strontium with lithium for the production of red light [45], or the use of high nitrogen species, but these are all still at the research stage. It also leads to related questions of performance (e.g. would a high-nitrogen-priming replacement be gun-fireable?) and again economic and environmental (are the complex molecular systems and salts manufacturable in an environmentally benign way that is affordable?). There has to be a systems justification for changing.

It must not be forgotten that some particular ingredients exist within a composition, not just for a pyrotechnic purpose but also to act in a different manner, e.g. the use of BaO_2 in a number of stores, because of its scavenging attributes, as well as its pyrotechnic properties. Will these materials need changing? What will happen?

There are a number of suggestions as to how we may move forward. There is a suggestion that to move forward, we must look to the past. How useful is gunpowder? How ubiquitous can it be? We know that from the sheer bulk of its use in the past that it can perform many tasks. Can natural products be reintroduced but with enhanced quality checks? Is there a cyclical nature to the understanding of materials? Will nano-sized materials enhance or detract from environmental issues? There are clearly problems with handling nanomaterials; however, it might be that nano-sized particles of more benign oxidizers and/or fuels will give performance that means they can replace the more noxious chemicals that are currently used. In addition, the emergence of novel, but as yet untried manufacturing methods may also lead to better, more controlled and less wasteful ways to work.

We need to have a systems thinking/approach; currently, it is estimated that 80% of all investigations within the industry are about how you turn a disparate set of objects into a system, not the chemistry of the pyrotechnics that it contains and thus creates polluting waste. So it is important to think of sequential follow through. What risk is posed for the next person in the chain?

This is a very generalized comment, but is just as applicable to the new materials that are coming from various studies. If their manufacturing is just as polluting as the methods for fabrication of current materials, is there any point in such a replacement? Without a systems viewpoint from raw materials acquisition through to end-of-life and engaging suppliers, manufacturers, and customers, then the development of more environmentally benign pyrotechnic materials will be pointless.

Manufacturing has duty of care from sourcing through to end-of-life/use, and currently the customer only considers use (so far). We need to think in a broader perspective; otherwise, given the fragmentation of the industry and service, any future moves to regulate will cause so much trouble it might not be worth continuing. It is only once a reasonable economic argument regarding the costs associated with waste and pollution and how reducing it will save companies money, will any real movement occur.

What is the real environmental cost of manufacturing? Should more be done to define input energy from renewable energy sources? What are the emissions associated with the manufacture of pyrotechnics? What is the real electricity usage and related carbon footprint for manufacturing plants, given that shipping is a major emitter of pollutants.

There is an alternative route, and that is to use non-pyrotechnic items where possible, for example, LED signalling flares [62], or the use of laser initiation of secondary explosives (removing the need for detonators) [63]. However, although this might appear to remove the issue of disposal of noxious chemicals, it merely delays it – how will an LED degrade compared to any residual tracer composition? Also, how benign are the raw materials sourcing, manufacture, and end-of-life disposal? It is not yet clear what will happen – even non-pyrotechnic items still have all the issues of item design and performance that pyrotechnic systems have and may have others, for example, in the case of initiators, using non-pyrotechnic methods may get electromagnetic interference in addition to toxicity and environmental concerns.

One last and pertinent viewpoint from the European perspective is whether the United States actually has more problems than the United Kingdom or EU, or have they simply studied it in more detail? To some degree, we all have to look after our own backyard and keep that clean and tidy – if we all do that, it makes looking after the whole thing much easier. Thinking globally, acting locally.

The drive to profit, stove-piping of knowledge, and cutting of communication ties between former colleagues in the various reorganizations worldwide has hampered general research in energetic systems and with the rise of environmental concerns has stymied investigations into this area. Pollution occurs and the materials that pyrotechnics use will have an effect on the local environment. Although the effect is primarily a local effect, e.g. at the site of manufacturing, usage, or disposal, this will have a broader effect and spread. One company or one nation causing an environmental impact has an effect on all.

There can also be a personal viewpoint. The blackthorn bush is a delightful source of sloes, which can be turned into fragrant jelly or delicious liqueur. The author once worked in a company whose borders were lined with blackthorn bushes, which always had plentiful fruit in the autumn, but the author was never once tempted to pick any. Can anyone who reads this chapter say they would pick nearby fruit and use it?

Acknowledgements

RV thanks a number of industrial colleagues for their time and patience with my attempts at interviews for this chapter. Alex Lay, Tracey Salt, and Owain Williams of Chemring, Tracey Vine of Qinetiq, Mark Bishop of Defence Science and Technology Laboratory (DSTL), and last, but certainly not least, Dave Holley of BAE Systems. Their thoughts have informed RV's thoughts, but any opinions are his alone and do not reflect any one individual's opinion or corporate viewpoint. RV would also like to show his appreciation to the Editors for being given the opportunity to be part of this work.

References

1 Sabatini, J.J. (2014). Advances towards the development of 'green' pyrotechnics. In: *Green Energetic Materials*, 1ee (ed. T. Brinck), 64–101. John Wiley & Sons.
2 Sabatini, J.J. (2018). A review of illuminating pyrotechnics. *Propellants Explos. Pyrotech.* 43: 28–37.
3 House of Commons Environmental Audit Committee The Future of Chemicals Regulation after the EU Referendum Eleventh Report of Session 2016–17 HC 912. https://www.parliament.uk/business/committees/committees-a-z/commons-select/environmental-audit-committee/inquiries/parliament-2015/inquiry2/ (accessed 26 June 2018).
4 House of Commons Environmental Audit Committee The Future of Chemicals Regulation after the EU Referendum: Government Response to the Committee's Eleventh Report of Session 2016–17 HC 313. https://www.parliament.uk/business/committees/committees-a-z/commons-select/environmental-audit-committee/inquiries/parliament-2015/inquiry2/ (accessed 26 June 2018).
5 Hartl, C., Konter, O., St. George, S. et al. (2018). Warfare dendrochronology – trees as witnesses of the Tirpitz attacks. *Geophys. Res. Abstr.* 20: 12769.
6 Davidovic-Milovanov, D. (1959). Menstrualni ciklus i ginekološki status radnica pirotehničke industrije. *Arhiv za higijenu rada i toksikologiju* 10: 43–52.
7 Bach, W., Dickinson, L., Weiner, B., and Costello, G. (1972). Some adverse health effects due to air pollution from fireworks. *Hawaii Med. J.* 31: 459–465.
8 Smith, R.M. and Dinh, V.-D. (1975). Changes in forced expiratory flow due to air pollution from fireworks. *Environ. Res.* 9: 321–331.
9 Bach, W., Daniels, A., Dickinson, L. et al. (1975). Fireworks pollution and health. *Int. J. Environ. Stud.* 7: 183–192.
10 Singh, A., Bloss, W.J., and Pope, F.D. (2015). Remember, remember the 5th of November; gunpowder, particles and smog. *Weather* 70: 320–324.

11 Sjögren, B., Ljunggren, K.G., Almkvist, O. et al. (1996). A follow-up study of five cases of aluminosis. *Int. Arch. Occup. Environ. Health* 68: 161–164.

12 Pope, R.J., Marshall, A.M., and O'Kane, B.O. (2016). Observing UK bonfire night pollution from space: analysis of atmospheric aerosol. *Weather* 71: 288–291.

13 Dyke, P., Coleman, P., and James, R. (1997). Dioxins in ambient air, bonfire night 1994. *Chemosphere* 34: 1191–1201.

14 Fleischer, O., Wichmann, H., and Lorenz, W. (1999). Release of polychlorinated dibenzo-*p*-dioxins and dibenzofurans by setting off fireworks. *Chemosphere* 39: 925–932.

15 Lee, R.G.M., Green, N.L.J., Lohmann, R., and Jones, K.C. (1999). Seasonal, anthropogenic, air mass, and meteorological influences on atmospheric concentrations of polychlorinated dibenzo-*p*-dioxins and dibenzofurans (PCDD/Fs): evidence for the diffuse combustion sources. *Environ. Sci. Technol.* 33: 2864–2871.

16 Schmidt, P., Bogdal, C., Wang, Z. et al. (2014). Releases of chlorobenzenes, chlorophenols and dioxins during fireworks. *Chemosphere* 114: 158–164.

17 Moreno, T., Querol, X., Alastuey, A. et al. (2007). Recreational atmospheric pollution episodes: inhalable metalliferous particles from firework displays. *Atmos. Environ.* 41: 913–922.

18 Moreno, T., Querol, X., Alastuey, A. et al. (2010). Effects of fireworks events on urban background trace metal aerosol concentrations: is the cocktail worth the show? *J. Hazard. Mater.* 183: 945–949.

19 Vecchi, R., Bernardoni, V., Cricchio, D. et al. (2008). The impact of fireworks on airborne particles. *Atmos. Environ.* 42: 1121–1132.

20 Godri, K.J., Green, D.C., Fuller, G.W. et al. (2010). Particulate oxidative burden associated with firework activity. *Environ. Sci. Technol.* 44: 8295–8301.

21 Steinhauser, G., Sterba, J.H., Foster, M. et al. (2008). Heavy metals from pyrotechnics in new years eve snow. *Atmos. Environ.* 42: 8616–8622.

22 Steinhauser, G. and Musilek, A. (2009). Do pyrotechnics contain radium? *Environ. Res. Lett.* 4: 034006.

23 Sterba, J.H., Steinhauser, G., and Grass, F. (2013). Illicit utilization of arsenic compounds in pyrotechnics? An analysis of the suspended particle emission during Vienna's new year fireworks. *J. Radioanal. Nucl. Chem.* 296: 237–243.

24 Camilleri, R. and Vella, A.J. (2010). Effect of fireworks on ambient air quality in Malta. *Atmos. Environ.* 44: 4521–4527.

25 Camilleri, R. and Vella, A.J. (2016). Emission factors for aerial pyrotechnics and use in assessing environmental impact of firework displays: case study from Malta. *Propellants Explos. Pyrotech.* 41: 273–280.

26 Vella, A.J., Chircop, C., Micallef, T., and Pace, C. (2015). Perchlorate in dust fall and indoor dust in Malta: an effect of fireworks. *Sci. Total Environ.* 521–522: 46–51.

27 Scheytt, T.J., Freywald, J., and Ptacek, C.J. (2011). Untersuchung ausgewählter oberflächen-, grund- und bodenwasserproben auf perchlorate in Deutschland: Erste ergebnisse. *Grundwasser – Z. der Fach. Hydrogeol.* 16: 37–43.

28 Joly, A., Smargiassi, A., Kosatsky, T. et al. (2010). Characterisation of particulate exposure during fireworks displays. *Atmos. Environ.* 44: 4325–4329.
29 Crespo, J., Yubero, E., Nicolàs, J.F. et al. (2012). High-time resolution and size-segregated elemental composition in high-intensity pyrotechnic exposures. *J. Hazard. Mater.* 241–242: 82–91.
30 Baranyai, E., Simon, E., Braun, M. et al. (2015). The effect of fireworks event on the amount and elemental concentration of deposited dust in the city of Debrecen, Hungary. *Air Qual. Atmos. Health* 8: 359–365.
31 Remškar, M., Tavčar, G., and Škapin, S.D. (2015). Sparklers as a nanohazard: size distribution measurements of the nanoparticles released from sparklers. *Air Qual. Atmos. Health* 8: 205–211.
32 Attri, A.K., Kumar, U., and Jain, V.K. (2001). Formation of ozone by fireworks. *Nature* 411: 1015.
33 Caballero, S., Galindo, N., Castañer, R. et al. (2015). Real-time measurements of ozone and UV radiation during pyrotechnic displays. *Aerosol Air Qual. Res.* 15: 2150–2157.
34 Ten Brink, H., Henzing, B., Otjes, R., and Weijers, E. (2018). Visibility in the Netherlands during new year's fireworks: the role of soot and salty aerosol products. *Atmos. Environ.* 173: 289–294.
35 Kok, L., Beng, O.T., and Lay, S.G. (1991). Simulation corrosion studies of the effect of red phosphorus with brass. In: *Proceedings of 17th International Pyrotechnics Seminar*. Beijing, China.
36 Strange, K., Mehay, S., Armstrong, C.D. and Hubble B.R. (2005). Phosphine measurement results taken during rework of M819 81 mm mortar cartridge (P148). *Proceedings of 36th Internationale ICT-Jahrestagung*, Karlsruhe, Germany.
37 Koch, E.-C. and Cudzilo, S. (2016). Safer pyrotechnic obscurants based on phosphorus(V) nitride. *Angew. Chem. Int. Ed.* 55: 15439–15442.
38 Miklaszewski, E.J., Shaw, A.P., Poret, J.C. et al. (2014). Performance and ageing of Mn/MnO_2 as an environmentally friendly energetic time delay composition. *ACS Sustainable Chem. Eng.* 2: 1312–1317.
39 Koenig, J.T., Shaw, A.P., Poret, J.C. et al. (2017). Performance of W/MnO_2 as an environmentally friendly energetic time delay composition. *ACS Sustainable Chem. Eng.* 5: 9477–9484.
40 Poret, J.C., Shaw, A.P., Csernica, C.M. et al. (2013). Versatile boron carbide-based energetic time delay compositions. *ACS Sustainable Chem. Eng.* 1: 1333–1338.
41 Shaw, A.P., Poret, J.C., Grau, H.A., and Gilbert, R.A. (2015). Demonstration of the $B_4C/NaIO_4/PTFE$ delay in the U.S. army hand-held signal. *ACS Sustainable Chem. Eng.* 3: 1558–1563.
42 Tichapondwa, S.M., Focke, W.W., Del Fabbro, O., and Kelly, C. (2015). Calcium sulphate as a possible oxidant in 'green' silicon-based pyrotechnic time delay compositions. *Propellants Explos. Pyrotech.* 40: 518–525.
43 Fischer, N., Klapötke, T.M., Marchner, S. et al. (2013). A selection of alkali and alkaline earth metal salts of 5,5'-bis(1-hyrdoxytetrazole) in pyrotechnic compositions. *Propellants Explos. Pyrotech.* 38: 448–459.

44 Klapötke, T.M., Müller, T.G., Rusan, M., and Stierstorfer, J. (2014). Metal salts of 4,5-dinitro-1,3-imidazole as colorants in pyrotechnic compositions. *Z. Anorg. Allg. Chem.* 640: 1347–1354.

45 Glück, J., Klapötke, T.M., Rusan, M. et al. (2017). A strontium and chlorine free pyrotechnic illuminant of high color purity. *Angew. Chem. Int. Ed.* 56: 16507–16509.

46 Fischer, N., Feller, M., Klapötke, T.M. et al. (2014). Spectroscopic investigations of high-nitrogen compounds for near-infrared illuminants. *Propellants Explos. Pyrotech.* 39: 166–172.

47 Moretti, J.D., Csernica, C.M., Poret, J.C. et al. (2017). Evaluation of a sustainable pyrotechnic tracer composition for small caliber ammunition. *ACS Sustainable Chem. Eng.* 5: 10657–10661.

48 Miklaszewski, E.J., Dilger, J.M., and Yamamoto, C.M. (2017). Development of a sustainable perchlorate-free yellow pyrotechnic signal flare. *ACS Sustainable Chem. Eng.* 5: 936–941.

49 Brusnahan, J.S., Shaw, A.P., Moretti, J.D., and Eck, W.S. (2017). Periodates as potential replacements for perchlorates in pyrotechnic compositions. *Propellants Explos. Pyrotech.* 42: 62–70.

50 Brusnahan, J.S., Poret, J.C., Moretti, J.D. et al. (2016). Use of magnesium diboride as a 'green' fuel for green illuminants. *ACS Sustainable Chem. Eng.* 4: 1827–1833.

51 Klapötke, T.M., Krumm, B., and Moll, R. (2013). Polynitroethyl- and fluorodinitroethyl substituted boron esters. *Chem. Eur. J.* 19: 12113–12123.

52 Klapötke, T.M., Rusan, M., and Sproll, V. (2013). Preparation of energetic poly(azolyl)borates as new environmentally benign green-light-emitting species for pyrotechnics. *Z. Anorg. Allg. Chem.* 639: 2433–2443.

53 Williams, M.R. and Vrcelj, R.M. (2018). Internal temperature and pressure and the rate of mass loss from hand thrown smoke grenades. *Propellants Explos. Pyrotech.* 43: 825–830.

54 Moretti, J.D., Sabatini, J.J., Shaw, A.P. et al. (2013). Prototype scale development of an environmentally benign yellow smoke hand-held signal formulation based on solvent yellow 33. *ACS Sustainable Chem. Eng.* 1: 673–678.

55 Moretti, J.D., Sabatini, J.J., Shaw, A.P., and Gilbert, R.A. (2014). Promising properties and system demonstration of an environmentally benign yellow smoke formulation for hand-held signals. *ACS Sustainable Chem. Eng.* 2: 1325–1330.

56 Shaw, A.P., Diviacchi, G., Black, E.L. et al. (2015). Versatile boron carbide-based visual obscurant compositions for smoke munitions. *ACS Sustainable Chem. Eng.* 3: 1248–1254.

57 Shaw, A.P., Poret, J.C., Gilbert, R.A. et al. (2013). Development and performance of boron carbide-based smoke compositions. *Propellants Explos. Pyrotech.* 38: 622–628.

58 Wilharm, C.K., Chin, A., and Pliskin, S.K. (2014). Thermochemical calculations for potassium ferrate(VI) K_2FeO_4 as a green oxidizer in pyrotechnic formulations. *Propellants Explos. Pyrotech.* 39: 173–179.

59 Koch, E.-C. Toxic combustion products. In: *Metal-Fluorocarbon Based Energetic Materials*, 326–333. Weinheim, Germany: Wiley.

60 Beach, A.B., Crews, R.C., Harrison, D.D. et al. (1998). Atmospheric dispersion of reacting chemicals from energetic materials. In: *4th Conference on Life Cycles of Energetic Materials*, 332. CA: Fullerton.
61 European Commission. For example, Skills in the defence sector. https://ec.europa.eu/growth/sectors/defence/industrial-policy/skills_en (accessed 26 June 2018).
62 Rothblum, A.M., Reubelt, V.A., and Lewandowski, M.J. (2015). Improving maritime distress signals: why a 130 cd flashing LED is more conspicuous than a 500 cd flare. In: *Proceedings of the Human Factors and Ergonomics Society 59th Annual Meeting*, 1696–1700. SAGE.
63 Yan, Z., Zhang, C., Liu, W. et al. (2016). Ultrafast laser-induced ignition of RDX single crystals. *Sci. Rep.* 6: 20251.

6

Munitions in the Sea

Sandro Carniel[1,2], Jacek Beldowski[3], and Margo Edwards[4]

[1] CNR-ISMAR, Castello 2737/F, 30122, Venice, Italy
[2] NATO STO CMRE, V.le San Bartolomeo 400, 19126, La Spezia, Italy
[3] Institute of Oceanology, Polish Academy of Sciences, Powstańców Warszawy 55, 81-712, Sopot, Poland
[4] Applied Research Lab, University of Hawaii, 2800 Woodlawn Dr, Honolulu, HI, 96822, USA

6.1 Introduction

In the years preceding World War I (WW I), and more so during and after World War II (WW II), many countries (i.e. the United States, the United Kingdom, Russia, Germany, Italy, France, and Canada) manufactured massive quantities of munitions that were filled with explosives or chemical agents. Some of these munitions, after they had been fused, were used at firing ranges or during military actions, but did not function as expected and remained in the environment – these munitions are known as unexploded ordnance (UXO). Other munitions were disposed of, usually without fuzes, to eliminate stockpiles of obsolete, excess, captured, or damaged munitions – these are known as disposed military munitions (DMMs). In combination, UXO, DMMs, and associated containers, bulk material, and other components are termed munitions and explosives of concern (MEC). Until the 1970s, a significant fraction of MEC was disposed of in the ocean, because the oceans were considered a safe and inexpensive solution for disposal of unwanted materials. Limited knowledge about the marine ecosystem resulted in the risks of MEC being accidentally collected by fishermen or problems caused by interactions between MEC and marine infrastructure to be considered negligible.

The production and disposal of MEC over the past few decades reflects the evolution of protocols and laws regarding these materials. Shortly after WW I, the Geneva Protocol (https://www.un.org/disarmament/wmd/bio/1925-geneva-protocol/) banned the use of chemical weapons in armed conflicts, thus initiating the first sanctioned dumping of chemical weapons in the 1920s. Sea disposal of 'substances that are liable to create hazards to human health', including chemical munitions or containers with chemical warfare agents (CWA) and conventional explosives, continued as a common practice until the Convention on the Prevention of Marine Pollution by Dumping of Wastes (Oslo Convention in

Energetic Materials and Munitions: Life Cycle Management, Environmental Impact and Demilitarization,
First Edition. Edited by Adam S. Cumming and Mark S. Johnson.
© 2019 Wiley-VCH Verlag GmbH & Co. KGaA. Published 2019 by Wiley-VCH Verlag GmbH & Co. KGaA.

1972) and other matter (also known as the London Convention) was enacted on 30 August 1975 [1].

In this chapter, we focus on general aspects related to MEC disposal at sea and conventional explosives.

Sea disposal of obsolete, excess, captured, or damaged MEC was performed by as many as 40 countries [2], mostly by the United States, France, the United Kingdom, Japan, and Russia. Different countries had different regulations regarding how disposal should be conducted. As an example, the United States required sea disposal to take place in marine waters of depth not less than 30 m and at a distance of minimum 65 km from the shore (the minimum depth was later increased to 55 m). However, these regulations were not shared with, nor adopted by, other countries, and MEC was frequently disposed of in coastal waters, sometimes where these were very shallow. Moreover, dumping was also conducted in freshwater environments such as lakes or rivers. Current estimates are that several million tonnes of MEC were dumped into waterways, seas, and oceans. There are 127 documented maritime areas where sea disposal is known, while over 300 total dumpsites are suspected to exist (James Martin Center, https://www.nonproliferation.org/). Dumpsites are located in the Atlantic, Pacific, and Indian Oceans, along the eastern and western coasts of Canada and the United States, in the Gulf of Mexico, near Australia, New Zealand, India, the Philippines, Japan, and the United Kingdom, and in the Caribbean, Mediterranean [3], Black, Red, Baltic, and North Seas, conferring on this problem the status of a global threat that demands internationally agreed protocols and actions to solve [4].

In Europe, North America, and southwest Pacific, the amount of conventional explosives sums up to millions of tonnes ([5]; Böttcher et al. [6] alone reports 1.6 million tonnes just in German North and Baltic Sea waters). Identified dumpsites of conventional and chemical munitions exist in the coastal waters of the North Sea (Belgium, Holland, Germany); deep sites are present in the Irish Sea and Biscay Bay, and selected sites in the semi-enclosed Adriatic and Baltic Seas [7]. North Sea sites originate from WW I, while the Baltic, Adriatic, and Irish Seas contain warfare material mostly from WW II [7, 8]. Altogether, the amount of chemical munitions in European Waters is estimated to be 684 000 tonnes [7].

A 2016 report to the United States Congress [9] documented that 27 000 tonnes of chemical warfare material had been disposed of in the waters around the United States, including Alaska, Hawaii, and US territories in the Caribbean.

Sea-disposed munitions represent a threat to commercial operations that interact with the sea bottom, for example, fishermen who trawl for their catch or energy companies that instal infrastructure for offshore resource extraction or export. MEC is also a potential point source of contaminants to benthic ecosystems. The highly variable and poorly understood corrosion of munitions casings directly affects dissolution and decomposition of explosives contained in the munitions, making it hard to predict the fate and transport of constituents released into the environment, as well as the temporal trends in the contaminant release and impact on the environment. Therefore, it is necessary to consider a variety of management options when dealing with dumping sites [10]. The simplest and most environmentally friendly option begins with an accurate monitoring of dumpsites, in order to assess and possibly affect the spread of contaminants. Such monitoring may include measuring met-oceanic conditions, levels of explosives and degradation products in the sediments and water column, environmental parameters responsible for water and suspended matter movements, corrosion rate of munitions, and the capability of toxic chemicals to transform. Oceanographic parameters such as turbulence, current speed and direction, and oxygen concentration are of particular interest in defining the environmental conditions hosting the munitions. Leaving alone blow-in-place, there are other options that can be adopted for managing dumpsites, although some of these may have adverse effects on the environment and are recommended only in cases where the impact of dumped munitions on the environment is well documented and exceeds threats created by a possible remediation activity.

Options alternative to monitoring dumpsites can be divided into two subgroups: *in situ* methods and remediation. *In situ methods* include hydrolysis of munitions constituents in underwater domes and various sediment capping options [10], which could either transform toxicants into less toxic compounds, or separate them from the bottom water for long enough to be buried by sediment layers or for the natural depuration processes to complete. The disadvantage of the *in situ* approach is the elimination of selected sea bottom areas from the ecosystem, in terms of habitat, and other ecosystem roles and services they normally provide. *In situ* methods were successfully demonstrated in the Black Sea, where post-Soviet chemical munitions were encased by concrete sediment capping [11].

Remediation includes various means of retrieval, although the most environmentally friendly option seems to be the *in situ* overpacking of partially corroded munitions into hermetic containers. Retrieval operations create risks of resuspension of contaminated sediments and the release of toxic agents into the water; however, such risks may be minimized by careful operation of divers and underwater robots. Retrieved munitions may then be neutralized using various means aboard ship or at land-based facilities. Neutralization methods include plasma ovens, detonation chambers, or combustion chambers, with specialized off-gassing treatment [8]. Because all of these options are costly, they should be preceded by a risk assessment and cost–benefit analyses. Recent examples of possible remediation are explored in the DAIMON (Decision Aid for Marine Munitions) project, funded by the European Union (EU) and aimed precisely at developing risk assessment methods and selecting remediation options.

6.2 The Controlling Factors

6.2.1 Environmental Aspects

Once dumped at sea, MEC are subject to the physical characteristics of the surrounding environment, which can be harsh compared to conditions on land. For example, objects disposed of in deep waters may implode as they descend to the seafloor if their casings are made of thin metal and air pockets are present within the casings interior, or the material is compressible resulting in a pressure differential across the munitions casings that creates cracks or crushes the casings. Shallow sites, where implosions are less frequent, may be subject to intense surface mixing and storms, which create higher velocity currents and pressure gradients that may transport munitions and/or resuspend and transport sediments that are contaminated by leaks from the munitions casings.

Munitions disposed of at sea may ultimately reside on a variety of bottom types perhaps composed of mud or clay or sand or rock. If the substrate is the same, the environment may not be, for example, if the region is characterized as erosional or depositional. Muddy bottoms are characterized by low-density sediments in which munitions may be buried upon encountering the seafloor. Investigations in the Gdańsk Deep, Baltic Sea, showed that munitions in the southern part of this region, where the density of sediments is low, were buried at depths from 1 to 2.5 m [12–14]. In such cases, degradation products are observed to spread within the sediments surrounding the objects, likely through the process of molecular diffusion. In the case of muddy sediments, the permeability of the sediments is generally so small that the process of dispersion caused by advection currents (either induced by density or resulting from the pressure gradient) can be disregarded.

Anthropogenic activities can play an important role in the redistribution of MEC. For example, trawling fishing nets have an important influence on seabed structure and bottom currents, and consequently modify advection and diffusion processes with respect to undisturbed conditions. Other anthropogenic effects include the deliberate dragging of MEC to new locations, for example, to make room for a power cable or pipeline or the re-disposal of munitions that have been accidentally pulled aboard a ship during operations.

6.2.2 Corrosion

As time progresses, metal objects resting on the seafloor rust and are subject to chemical and mechanical erosion. At some point, metal munitions casings will corrode to the point that there is direct contact between seawater and the munitions constituents. At this point, the effects will depend on several factors. As an example, although some studies assessed plausible rates of bombshell corrosion in the range of 0.05–0.575 mm/year, it was found that real corrosion can be increased up to fourfold by moderate stirring/current of the seawater [15], both in the effect of water exchange, leading to more effective diffusion, and the mechanical erosion in shallow places caused by transport of sand particles along the surface of munitions [16]. As further evidence of the complexity of shell corrosion as a function of the specific type, case, and manufacturing, the Hawaii Undersea Military Munitions Assessment (HUMMA) documented the existence of trails of munitions that were disposed of in the waters south of Oahu, Hawaii [17]. Visual inspection of these trails using human-occupied vehicles (HOVs), and remotely operated vehicles (ROVs) revealed that individual trails were composed almost entirely of only one or two types of munitions [18]. This observation was consistent with the storage and disposal procedures of the mid-twentieth century, which directed personnel to keep munitions of a similar type together for storage and subsequent disposal. Despite the consistency of the type of MEC within individual trails, the munitions conditions were highly variable, with highly corroded bombs frequently observed next to others of the same kind that appeared to be only moderately weathered.

Saltation can produce mechanical erosion, accelerating the corrosion processes affecting munitions casings. Once casings are breached, the internal contents can be dispersed along the main hydrodynamic regime into the surrounding environment. Sometimes advection processes are dominant, associated with turbulent diffusion, and affecting the water and sediments in the immediate vicinity of the point source. Other times molecular diffusion dominates, slowly redistributing the internal constituents, usually over a smaller area. Other processes that may affect the fate and transport of materiel contained within munitions that are buried in sediments include the following:

(i) *Settling sediments*: The increasing weight of the growing and settling of the topmost sediment layer results in the expulsion of water from deeper sediment layers in a process called compaction, which might disperse contaminants.
(ii) *Rising gas*: Fermentation gases generated by anaerobic biological degradation can form small bubbles that rise buoyantly and cause micro-turbulence, resulting in constituents being ejected into the seabed or water column.
(iii) *Bioturbation:* Several species of biota can affect the mobilization of polluted sediments via habitation within the seabed or regular visits to feed on benthic organisms. When animals burrow into the seafloor or scour the sediment for prey, they can mix and release the layers of particles into the near-bottom water. Resuspended sediments can thus be dispersed via near-bottom water movements.

In addition to saltation, factors that act as catalysts to corrode munitions casings include temperature and oxidization. Higher temperatures and oxygen content in sediments or bottom water tend to increase the rate of corrosion of the metal containers; in contrast, chemically reducing conditions with low temperatures and oxygen content contribute to the persistence of sea-disposed munitions [19]. Interactions with benthic biota can also cause chemical or mechanical erosion depending on how animals attach themselves to munitions casings or how frequently they physically interact with munitions casings. For example, at the HUMMA deep-water site in Hawaii, sea stars have been observed to migrate across munitions [18] and shrimp and rays have been observed to stir up materials inside and outside of munitions casings (www.hummaproject.com).

The casings and driving bands of munitions were often produced using different metal types (steel, copper, etc.), making them susceptible to galvanic corrosion. Biological overgrowth may reduce or further influence corrosion rates [20, 21].

Hydrodynamic conditions of the area where munitions have been dumped can heavily affect corrosion rates as well; energetic regions characterized by strong current speeds or wave action [22] correlate well with higher corrosion rates.

6.2.3 Fate and Transport of Constituents

Both parent substances and chemical degradation products resulting from MEC can be transported by seawater; in most cases, they are first adsorbed by sediments and transported with moving silt or clay if the bottom currents in the area are sufficiently strong and the seafloor type and relief do not trap the sediments. Moving water masses induce shear stress on the bottom, and when it exceeds a critical value, which depends on bottom characteristics such as grain size and roughness bedload movement, is ignited. This movement can result in resuspension or saltation, especially of small particles, which may be moved away by currents. Several model scenarios have been developed to predict spreading of chemicals released from munitions, both in terms of Lagrangian transportation of substances diluted in the water (plumes) and in association with sediment grains. The main drivers that influence the dispersal of munitions constituents are the existing hydrodynamic forces within the disposal area.

Although thorough assessments have been conducted on a relatively limited amount of cases, to date there is little evidence that toxic chemicals from munitions can be found in high concentrations in seawater, except very close to the munition source [23]. Many harmful chemical compounds are known to experience hydrolysis that rapidly transform to non-toxic or low-toxic states, but in case of TNT and its degradation products, ADNT and DANT, the latter are more toxic than the parent compound. In case of dissolution, further hydrolysis and decomposition may remove those substances from the water column; while in the case of less reactive compounds, dilution prevails, as they will be transported as particulates mostly and removed from water by sedimentation.

Although there is a possibility that more toxic compounds can be released as munitions casings that have remained intact begin to lose containment, given the pronounced variability in the integrity of munitions casings, even with those of the same type [18, 24], it seems very unlikely that there could be a synchronized release of munitions constituents on a broad scale.

Theoretical studies on munitions compounds' migration in sediments after release from buried munitions were performed by Francken and Hafez [25] for a North Sea region, to be used for objects buried below 1 m under the sediment/water interface. According to the models employed, the extent of sediment pollution can extend from a few centimetres to several metres around the objects (possibly depending on whether the MC are released into the water column or porewater). However, it may well be that the objects did not sink below the sediment/water interface upon dumping, and that they rest on harder sediments; in many cases, even if they were dropped in the area of sediment accumulation, the rate of natural sedimentation is not capable of separating them effectively from the surrounding environment. In the Baltic region, for instance, the low sedimentation rate [26] can provide just 15 cm within 70 years, which may surely slow down the corrosion but not stop the diffusion. Moreover, hydrophobic particles of warfare agents may attach to topmost, unconsolidated sediments and be transported to other areas in resuspension events. Such a process was observed even in sandy areas, which are periodically covered by fine particles forming a so-called fluffy layer of suspended matter (FLSM) [27].

In any case, the community agrees that there exists a strong need to increase the analysis of possible contaminant distribution for a larger number of cases, and assess what could be acute polluting processes in the vicinity of large dumping sites.

6.2.4 Sea-Disposal Process

The disposal process itself is another important factor to consider when searching for munitions at sea and assessing the degree of their corrosion. For example, munitions disposed of in boxes were observed to float before sinking, allowing them to remain longer in the water before leaking contaminants and allowing the currents to disperse them more widely before they finally sank to the seafloor. In contrast, munitions disposed without any containerization were dense and sank quickly and without much lateral movement, remaining closer to the location where they were pushed over the side. In historical documents from the United States, there are reports of munitions being shot at during deployment in order to make then sink more rapidly. Any bullet that perforated a munitions casing created near-instantaneous contact between the interior and the seawater, while inhibiting other processes (e.g. implosion).

6.3 Tools for Assessment and Remediation

Assessing the risk posed by MEC is a critical first step at disposal sites. A variety of technologies are available to detect and assess munitions, and selection of the approach should be based on the environmental conditions of the area, the types of MEC thought to be in the area, and how munitions might rest, or be buried, on the seafloor. Most assessments of sea-disposal sites use a nested approach, which begins with the general reconnaissance of the site to understand topography, substrate, and water properties and to search for objects that appear man-made and other potential hazards. Subsequent efforts focus on small areas of interest at higher (and higher) resolution, identifying locations with potential targets. The most effective assessments typically involve multiple surveys and techniques.

Typically, three types of geophysical tools are used to assess sea-disposal sites: acoustic, electromagnetic (EM), and optical. Acoustic systems such as side-scan sonars and sub-bottom penetrators measure the response of objects to sound waves; they are useful for regional mapping of the geology and for identifying individual targets in settings where metallic objects have an acoustic response different from that of the surrounding seafloor. EM sensors measure the EM field of objects relative to the surrounding environment. Optical sensors such as cameras and laser scanners depict how objects on the seafloor respond to light.

6.3.1 Acoustic Sensors

Visible light is absorbed within 100 m of the ocean surface even in the clearest waters. In contrast, sound waves can penetrate to the deepest part of the ocean and into the sub-seafloor if they have long wavelengths and sufficient power. Any investigation to assess sea-disposal sites thus benefits from the use of acoustic sensors, which measure seafloor shape, seafloor reflectivity, and impedance changes in the sub-seafloor. Seafloor shape, more commonly known as bathymetry, represents underwater elevation changes. Seafloor reflectivity represents the smoothness/roughness of the seafloor and can be used to find objects on or just below the seafloor that have high acoustic contrast relative to their surroundings. Acoustic impedance contrasts in the sub-seafloor can represent changes in the vertical stratigraphy of the seafloor or detect objects with acoustic properties different from the materials in which they are buried.

6.3.2 EM Sensors

The majority of munitions casings were fabricated from ferromagnetic materials (e.g. steel) that exert local effects on the earth's magnetic field. As a result, precise measurement of the magnetic field will measure the contributions from metallic objects that appear as anomalies relative to the surrounding environment. Two types of EM sensors are used for mapping EM fields: electromagnetic induction (EMI) systems and magnetometers. EMI systems are usually active sensors, while magnetometers are passive, such that magnetometers can only detect ferrous metal, while EMI sensors detect both ferrous and non-ferrous metals. Magnetometers measure irregularities in the earth's magnetic field caused by the ferromagnetic materials such as those in munitions and are typically more effective for locating large, deep, ferrous objects. EMI systems induce an EM field and measure the response of objects near the sensor. EMI systems measure the secondary magnetic field induced in metal objects.

6.3.3 Optical Sensors

Optical sensors include still image or video camera equipment used to photograph the seafloor. They are usually accompanied by an artificial light source when assessing the seafloor because cameras require that the objects be exposed on the seafloor to be measured, which in turn depends on light and turbidity. As light decreases or turbidity increases, the effective field of view is reduced. Laser line scanners (LLSs) use lasers to illuminate and record detailed video over areas that can be significantly larger than the areas imaged by conventional underwater cameras. If water clarity is good, LLSs can measure objects at the scale of centimetres over areas of tens of metres. LLSs work by generating laser pulses that generate line scans of the bottom. These scans are captured by a high-speed camera, which combines them to make a swath of imagery that continuously expands while the system operates.

6.3.4 Platforms

All of the sensors described are mounted on some type of platforms to assess the sea-disposal sites. Platforms include surface vessels, towed vehicles and ROVs, and unmanned vehicles and autonomous underwater vehicles (AUVs. Selecting a suitable platform for each nested survey depends on the ability of the platform to operate effectively; for example, a surface vessel is better suited for conducting a large-area reconnaissance survey than a near-bottom-operating AUV.

MEC surveys require the use of watercraft for mapping or towing or deploying other platforms. The cost of using a vessel for a project can be significant, and emphasis should be placed on using a vessel capable of performing the operations under average environmental conditions for the site. Consideration for adequate vessels should include the amount of deck space available for equipment, whether there are sufficient berths for the survey team, how long the vessel can remain at sea and under what conditions, and how much power the vessel can provide for ancillary systems. In addition, some over-the-side equipment requires that ships be equipped with winches and dynamic positioning systems. Watercraft have advantages over underwater platforms that may include the ability to communicate with the shore, take advantage of satellite navigation systems, and dynamically position themselves in specific locations.

Unmanned surface vehicles (USVs) are surface vessels that are operated remotely and/or follow the commands preprogrammed into their control systems. USVs are typically smaller than surface vessels, and are powered by rechargeable batteries, using the air around them to combust fuel, or by harvesting solar or wave energy. Like surface vessels, USVs can avail themselves of satellite navigation and communicate with operators who are not in their immediate vicinity.

Two types of ROVs are well-suited for conducting surveys: (i) ROVs equipped with thrusters for maneuvering and (ii) remotely operated towed vehicles. The former are best suited to work in small, well-defined areas in search of specific targets or for surveying a narrowly defined area such as a pipeline survey or cable route. Remotely operated towed vehicles can cover more ground effectively, and the addition of hydrodynamic surfaces and control systems allows the vehicle to automatically adjust its path to accomplish survey objectives such as operating at a constant altitude above the bottom.

AUVs range in size from small, cylindrical objects able to be deployed and recovered by a single person to vehicles many metres in length, weighing several tons, and requiring large ships to accommodate their launch and recovery systems. AUVs are typically equipped with systems for power, propulsion, command-and-control, navigation, and acoustic telemetry, in addition to a payload section that supports a variety of sensors. Acoustic telemetry for AUVs usually has limited bandwidth and range, so data tends to be stored onboard, with only small packets of information transmitted during operations. Position and command-and-control parameters are usually transmitted continuously, while sensor data, if they are transmitted at all, are subsampled to reduce data volume.

6.3.5 Navigation and Positioning

Without accurate and precise information about the location of the data collected, MEC surveys are not useful, so navigation and positioning of platforms and sensors are critical components in any underwater investigation. It is best to use the most accurate and cost-effective positioning systems available, although care must be taken in selecting how positioning systems are used and relate to one another. Many navigation and positioning systems can be combined with underwater platforms to provide positional accuracy on the order of a metre or even better.

The Global Navigation Satellite System (GNSS) provides positioning information using radio signals from a constellation of orbiting satellites. The original GNSS system was the United States Global Positioning System (GPS), which was followed by the Russian GLObalnaya NAvigatsionnaya Sputnikovaya Sistema (GLONASS), the European Union's Galileo, and China's BeiDou. Today's GNSS receivers using GPS or GLONASS individually have positional accuracy of 5–10 m; receivers combining both systems improve accuracy to ±2 m. DGPS/RTK-GPS has wide usage in underwater surveying. DGPS/RTK-GPS receivers are mounted to fixed points (e.g. on a survey vessel or a pier) to provide an accurate surface position.

The radio frequencies used by GNSS are absorbed by water, so determining a platform's position underwater is typically undertaken using sound waves (acoustics). The speed of sound in water is approximately constant, so by measuring the travel time for a sound wave to travel between two points, the distance between the points can be calculated. Three common acoustic systems are used to determine positions underwater: long baseline (LBL), short baseline (SBL), and ultrashort baseline (USBL). The information from these systems can be integrated with positions derived from GNSS above the water to provide an underwater position co-registered to a surface GNSS location. The LBL positioning system relies on multiple acoustic transponders positioned on the seafloor that are interrogated by an acoustic signal from a platform; each transponder replies with a unique acoustic signal and the position of the source is calculated on the basis of the known positions of the transponders and the two-way travel time to each. SBL and USBL underwater positioning systems rely on transponders mounted on the surface vessel rather than on the seafloor. For SBL systems, three to four transponders are mounted to the vessel as far apart as possible, and ranging is done between those transponders and a responder on an underwater platform. For USBL systems, multiple transducers are built into a single transceiver mounted on the support vessel; the system determines a range to the underwater transponder through travel time like the SBL system, but then measures the phase difference between when an echo is received at the hydrophones and calculating the azimuth and elevation angles.

Inertial navigation systems (INSs) can also be used in combination with a Doppler velocity log (DVL) and depth sensor, all integrated with GNSS, for AUV positioning. AUVs typically carry a GNSS receiver, but it only functions when the platform is at the ocean surface. Once the AUV submerges, the DVL bounces sound waves off the seafloor and by measuring the Doppler shift of the reflected waves computes the platform velocity over the seafloor. The INS measures accelerations, rotations, and magnetic field to determine heading to supplement the velocity data with platform orientation. INSs tend to be highly accurate initially, but they lose accuracy due to the inherent drift in the inertial heading sensors as distance and time increase since the last contact with GNSS. The use of INSs requires an AUV to surface regularly to maintain accurate positioning information, which increases the time to complete a survey.

6.3.6 Remediation

To date, remediation efforts for sea-disposed MEC have focused on testing and evaluating differing approaches; there are not yet examples of large-scale remediation efforts that employ one particular type of technology to remove most of the MEC from a site. Methods for remediating MEC can be grouped into two general categories (i) *in situ*, including blow-in-place and (ii) recover.

The most common approaches for remediating MEC in place involve detonation or inflagration using donor charge and encasing the munitions with a protective layer of a durable substance, for example, concrete.

The most common clearance method of dumped conventional munition is *in situ* detonation with a donor charge. Such clearance is usually assisted by mitigation strategy for protection of marine biota, but those are limited in many cases to noise mitigation measures, as well as detonation of a scaring charge, to remove sensitive life from the vicinity.

This practice is commonly adopted by many naval forces worldwide, which does not necessarily mean that it is environmentally friendly.

During *in situ* detonation, toxins still contained in munitions and those included in nearby sediments are both delivered to the water column and may contaminate large areas, and could be ultimately incorporated by marine organisms. Earlier studies show that during blasting, significant TNT and RDX residues are observed in the blasting site [28]. Especially, low-order detonations preferred for clearance can distribute large pieces of UXO in the environment [29].

For chemical munitions, the drilling approach could be used. The approach involves creating a small-diameter hole in munitions casings and using suction to draw the munitions constituents into a container that can be safely recovered at the ocean surface or on shore. An advantage of this approach is that it can be environmentally safer; for example, it would have less impact on biota that had attached itself to the munitions casings in a coral reef. The disadvantages are that this approach can be time-consuming and that there is the potential for leakage of toxins once the materiel has been removed from inside munitions casings. Efforts are undertaken to adapt a drilling approach for conventional munitions, using water jets to break solid explosives into particles small enough to suction (ROBEMM project, http://www.munitionsraeumung-meer.de/en/national-research/robemm/). Encasing MEC in concrete can be a faster and more cost-effective solution; but depending on the distribution of munitions and nearby animals, it can have a significant, negative impact on the environment.

Recovery of MEC involves the extraction of the entire munitions from the seafloor, which can have a significant environmental impact if a local ecosystem has incorporated the MEC into the habitat, and can be dangerous if the munitions casing integrity has deteriorated to the point of collapse, and even more if the MEC have been fused. Like recovery of munitions constituents from inside a casing, transport of munitions from the seafloor to a ship and/or back to shore also introduces numerous opportunities for explosion or accidental spillage that will affect equipment and personnel.

6.4 The Outstanding Problems

6.4.1 Technical Aspects

6.4.1.1 Location

As has emerged from the consideration of these sections, once munitions are dumped at sea without accurate tracing, they are extremely difficult to find. Moreover, even though major dumpsites positions are well-known, location of individual MEC is usually a difficult exercise.

This is even more valid for dumped munitions in periods when navigation was not precise, ship logs were not kept, etc. Within the CHEMSEA (Chemical Munitions Search and Assessment, 2011–2014) project, studies performed in Gotland Deep in the Baltic revealed the existence of 40 000 contacts on side-scan sonar in a dumping area of $1500\,km^2$. After careful categorization by automated and operator-assisted procedures, roughly 18 000 were ranked as probable munitions. Interestingly enough, these contacts were scattered across the dumpsite area, with many even outside (similar to the Skagerrak dumpsite -see [30]). Further evaluation was performed by means of ROV examination (these missions being very demanding, only 200 were performed) confirmed that 40% of designated targets were munitions [31].

This example points out the difficulty in locating munitions once they have not been properly identified at the beginning of the dumping. It is an effort very demanding in terms of shiptime, and one that can be performed only in calm weather (wave height not exceeding 1.5 m). Searching for munitions is still an outstanding problem when dealing with MEC at sea and is a type of artisanal knowledge that forces scientists to use many different tools from the toolbox.

6.4.1.2 Detection

Detection of dumped munitions, except when they are buried, relies mostly on acoustic methods of bottom mapping. Side-scan and synthetic aperture sonars are most common methods, although multi-beam echo sounders can also be used. In these cases, however, munitions have to be on the surface of sediments, and their contours visible. Such searches produce a large series of 'false positives', which have to be further verified.

Another possible method is magnetometry, which enables detection of metal targets, but turns out to be useless in case of completely corroded munitions.

In case of objects covered with sediments, sub-bottom acoustic profilers are then used. All those methods have limited range, since swath size is inversely proportional to resolution. In recent years, the usage of autonomous platforms was introduced, which in theory could reduce the shiptime since AUVs can operate independently of the ship, freeing it for other tasks. In practice, the safety of the vehicle is the new constraint, and the main ship has to stay in close vicinity of the operating AUV, in order to take action in case of malfunction.

Another issue is identification of munitions. In many areas, water visibility is poor, because of suspended matter close to the seafloor. In such cases, video searches are augmented by acoustic cameras or front-looking sonars, which helps in pinpointing target location.

In the Baltic Sea, visibility is sometimes reduced to 20 cm in dumpsites, which makes the identification of munitions very difficult. In such cases, the front-looking sonar is used to recognize the overall shape of the object, followed by visual identification from close range. Even then, it is difficult to discern chemical from conventional munitions.

Final identification is conducted by chemical methods. There are very few experimental *in situ* sensors, so it is mostly based on retrieving sediment sample and analysis in the laboratory. Some methods allow on-board detection of some agents and their degradation products [12].

6.4.1.3 Monitoring

Monitoring of dumped munitions is based mostly on evaluating corrosion progress and leakage into the marine environment. Due to the limitations described, it is usually based on single, well-known, and positioned objects. This, of course, raises the problem of sinopticity, since the local conditions are not always representative of a larger area, not to mention the whole dumping region. Hopefully, detection of MC in the water column would help constrain the release and regional spread of the contaminants. Monitoring can then be supplemented by observation of biota (bioaccumulation of chemicals, biodiversity, and biomarkers and effects of environmental stress on health). Nevertheless, since in many dumpsites due to the depth and oxygen deficiency the bottom life is limited, the usage of common biomonitoring organisms, macrozoobenthos, like bivalves and crustaceans, is impossible.

Hence biomonitoring is still carried out, but limited to biodiversity of infauna, e.g. nematodes, which are resilient enough to survive in the harsh environment, and fish studies [32, 33]. Fish studies are based on general conditions, prevalence of diseases, and parasites. One possible limitation to this kind of studies is that fishes are not sedentary, so results need to be accounted for the fact that populations may have spent most of their lifetime in different regions [13].

6.4.1.4 Handling

Unlike on land, the underwater environment usually restricts access to, and has a significant impact on, the performance of operators trained to work with MEC. This is true both in the case of shallow water operations, where direct interventions are possible, and in deep regions, where ROVs are usually sent and operated at distance.

The use of robotics and remotely operated equipment can clearly minimize the hazard to people; moreover, it brings in another advantage, since it allows work around the clock even at great operation depths.

The field of an underwater manipulator system is therefore a very promising one, and it can even keep Navy divers out of water. In order to be effective, such equipment have to deal with underwater environmental conditions, possible direct current drag on the system, and, of course, precision movements. In this way, they can supplement any direct operation on the munitions (blasting, icing with liquid nitrogen, jet cutting).

6.4.2 Environmental Aspects

Risks from munitions in the underwater environment are linked both to explosive and environmental hazards, associated with the release of the constituents contained in the MEC. We do not discuss here the exposure to direct explosion, and focus more on possible environmental impacts.

6.4.2.1 Chemical Degradation of MEC

The environmental fate of a MEC once in marine waters is an outstanding problem depending on many factors. The attitude of undergoing chemical transformations and the final pathways of environmental distribution are functions of the compounds (e.g. reactivity, polarity) and of the prevailing ambient conditions (e.g. temperature, salinity, hydrodynamics, bacterial population).

Chemicals resulting from these complex transformations may have close or different properties with respect to the 'parent' compounds.

Generally speaking, chemicals that degrade rapidly will tend to be less persistent in the environment. Chemicals that transform very slowly are the most environmentally persistent species and, given suitable hydrophobic (fat-soluble) properties, have the potential to bioaccumulate in living organisms via food webs (food chains). Bioaccumulation of these MC from aqueous exposure and ingestion in benthic invertebrates and fish appears to be low, and depuration rates relatively high, with half-life of the order of hours) [34–37]. In contrast, TNT transformation products do tend to bioaccumulate to a greater extent, and are depurated more slowly [34], and persistent organic pollutants (POPs) are one of the principal issues of environmental pollution.

Although there exist other explosive compounds used historically or which have been more recently developed, the larger amount of conventional munition materials present in terrestrial and marine environment [38] is represented by the following three types: the nitroaromatic 2,4,6-trinitrotoluene (TNT), the nitramines hexahydro-1,3,5-trinitro-1,3,5-triazine (RDX, or hexogen) and octahydro-1,3,5,7-tetranitro-1,3,5,7-tetreazocine (HMX, or octogen). We then devote particular attention to TNT and RDX and HMX, present as RDX by-product until the 1950s, when it was then used directly [5].

TNT is the most common type of explosive found in dumped munitions, frequently mixed with other energetic compounds (e.g. RDX). Explosives such as TNT are relatively stable, and detonation is induced by a primary explosive, or a so-called initiator. The most common primary explosives in use currently are fulminates, styphnates, and azides of mercury, lead, and silver [39].

TNT is characterized by a low solubility (see subsequent text); and during the environmental degradation, TNT DNO isomers (2,4-DNT and 2,6-DNT) may be formed. TNT may be also transformed to 2-amino-4,6-dinitrotoluene (2-A-4,6-DNT) and 4-amino-2,6-dinitrotoluene (4-A-2,6-DNT) due to biotic transformation of nitro groups to amino groups.

Although amino dinitrotoluenes have a low octanol–water partitioning coefficient, they may bind covalently to organic and mineral components in soil and sediment.

RDX, occurring in a variety of explosives, is also highly stable with poorer solubility than TNT, and has an octanol–water partitioning coefficient of log $K_{ow} = 0.86$, indicating that after dissolution it will not sorb readily into sediments.

The environmental fate and potential hazard of energetic compounds in the environment is affected by a number of processes including dissolution, sorption, abiotic and biotic transformation/degradation, and bioaccumulation [28].

After dissolution, explosives may be absorbed to minerals, amorphous grain coatings of metal oxyhydroxide, humic material, organic/inorganic colloids, or microorganisms. Existing studies are mostly based on soil experiments, showing reversible sorption for TNT and almost irreversible for RDX, showing a linear relationship with organic carbon content [40].

Hydrolysis, which is usually the fastest process of decomposition, is limited in these explosives, as nitroaromatics and aromatic amines are generally resistant to hydrolysis. Explosives containing a nitro functional group can undergo abiotic reduction, when nitro groups are converted to amino groups – this process is sensitive to pH and redox potential [41].

The rate of reduction can be further accelerated by microorganisms, but it is generally believed that abiotic process requires a catalyst – it could be iron compounds, clay minerals, and organic matter – which are highly abundant in sediments. Both RDX and TNT reduction were observed in the presence of ferrous ion and organic matter accompanied by H_2S [42]. Microbial transformation of energetic compounds results in a number of reactions including polymerization, covalent binding, and complexation [29].

Modeling efforts suggest that TNT release from submerged munitions is likely to be slow, with predicted concentrations in water near the munitions only of the order of nanograms per litre [43].

6.4.2.2 Long-Term and Long-Distance Transport

Dissolved dangerous material could be spread by natural and anthropogenic processes.

The main drivers that will have an influence on scattering of the MC pollution are hydrodynamic forces in the dumpsite area. However, it could be disturbed by anthropogenic activities. While it is not common that the munitions are moved from one place to another (however, it is possible, for example, to drag the shells or containers), there exist several anthropogenic processes that could have some influence on natural processes. For example, trawling fishing nets could have an important influence on seabed and bottom current structure, and impact advection/diffusion processes compared to undisturbed conditions.

Natural processes that could have an influence on spread of pollution are well-known (however, the influence of each one of them could be difficult to estimate). They include several different processes of differing strength. Permanent, low-force processes, which involve diffusion from sources, result in contamination of adjacent sediments, porewater, and water in the immediate vicinity of the leaking chemical munitions. Low to medium forces – which are frequent in the dumpsites – are horizontal currents of ordinary magnitude and eddy diffusion; disturbance by biota (bioturbation); and vertical transportation with bio-generated gas in the sediment and porewater from deeper layers, squeezed out due to the increasing weight of settling particles. In addition to horizontal movements, the water also undergoes vertical mixing. Observation in Baltic dumpsites [44] shows that other noteworthy near-bottom turbulences occur in deeper waters as well. Occasionally, events of stronger force, like the inflow of cold, salty, and oxygenated water from the North Sea into deep basins of the Baltic Sea and strong currents caused by storm surges and ice (in more shallow or coastal waters), may lead to sediment mass erosion and resuspension.

During the MODUM (Monitoring of Dumped Munitions, 2013–2016) project, two independent methods for estimation of potential leakage were developed. The first one is Lagrangian tracking of particles with random disturbance; the second uses a passive tracer as a marker of potential leakage. According to those models, during average Baltic currents, not exceeding 40 cm/s, maximum travel distance of particles originating at the dumpsites (at c. 100 m depth) varied from 20 to 75 km. These estimates are clearly very much site-dependent, and would turn out to be extremely valuable in other dumping sites [12].

6.4.2.3 Ecotoxicological Aspects

When it comes to discussing the effective consequences related to the MC exposure at sea, a critical factor controlling the environmental processes is related to the compounds' dissolution.

There exist a substantial variability in measured solubility at temperatures relevant to the marine environment. For example, at 20 °C, reported solubility for TNT in seawater varies between 70 and 120 mg/l amongst different studies. The effect of pH on TNT dissolution seems to be minimal, at least when remaining within the environmentally relevant temperature ranges [45, 46].

On the other hand, munition dissolution in seawater seems to depend mostly on physical processes regulating transport in the investigated dumping area, and the associated release may therefore vary significantly among sites (e.g. depending on wave currents). This is consistent with the results of Porter et al. [23] who showed that dissolved TNT was saturated within breached munitions, but concentrations declined by over three orders of magnitude within 10 cm of the munition surface.

Solubility curves for RDX and HMX show 3- and 60-fold lower solubility than those for TNT, presenting an exponential increase with temperature as well, although less data is available for these compounds with respect to TNT.

Among the toxic materials contained in the conventional explosives at sea, the most hazardous organic and inorganic compounds of conventional ammunition are TNT and its degradation products (trinitrobenzene, TNB; aminodinitrotoluene, ADNT) as well as mercury (Hg), methyl mercury that can be formed from the Hg release (MeHg), and lead (Pb), which are used in trigger units and explosive capsules [47]. These substances can be released into the seawater and the surrounding sediment also because of the corrosion of the metal housings of the weapons, which contaminates the nearby area.

It is indeed worth remembering that conventional munitions also bring along an associated quantity of other metals in their casings and fuses, among which are lead, iron, aluminium, copper, and zinc [48, 49].

Studies in marine environment show evidence of sediment contamination in the vicinity of dumped munitions by trace metals, in case of different types of munitions, both chemical and conventional [31, 50, 51].

However, although underwater conventional munitions do contain toxic metals such as lead or mercury, the chemical contamination associated with their presence and degradation appears almost exclusively to be organic energetic compounds.

It should also be noted that current background heavy metal concentrations make it difficult to clearly trace back their presence to a specific munition source [5]. This problem is also made more evident by the lack of specific studies and sampling strategies in this direction. All in all, although not being the primary source of concern, heavy metals may clearly add an additional negative ecological impact on the marine biological communities already exposed to organic energetic compounds.

In the few studies where sound analyses have been conducted on water samples collected very close to munitions, or even within breached ones, it is shown that dissolved munitions compounds' concentrations approach the solubility limits (mg/l range). Generally speaking, dissolved MCs are rapidly diluted away from the munition point source, although it has to be noted that analytical detection limits have largely prevented accurate quantification at levels present in the free water column or sediments (typically, we are dealing with ng/l or µg/kg). In addition, while MC dissolution rates from solid explosives have been investigated under controlled experimental conditions, the obtained information do not always represent the characteristic conditions experienced by underwater munitions [5], where observations indicate slower dissolution rates. For this reason, a larger number of *in situ* measurements are required and would surely provide more realistic rates, although such operations can be easily hampered by the hazards represented by working with explosives in real conditions.

In humans, TNT is associated with abnormal liver function and anaemia, and both TNT and RDX have been classified as potential human carcinogens. In general, according to toxicity studies, acute toxicity due to MC levels on large marine bottom biota areas seems to be rather unlikely.

However, there exist studies with *Vibrio fischeri* that have established TNT as being 'very toxic' to aquatic organisms [47], mutagenic, with some metabolites more so than the TNT itself. TNT and some of its metabolites display also chronic toxicity in sediments, showing the ability to bioaccumulate in benthic biota, and lethal effects [52]. Also, degradation products of TNT, such as 2,4-diaminonitrotoluene (2,4-DANT) and 1,3,5- trinitrobenzene (TNB), display chronic toxicity, resulting in lethal effects, and in case of 2,4-DANT also significant growth reduction [53].

Studies performed by Nipper [54] showed that zoobenthos can be locally affected severely by tetryl and 1,3,5-trinitrobenzene, whereas among the dinitro- and trinitrotoluenes and benzenes, toxicity tended to increase with the level of nitrogenation.

All in all, strong *caveats* need to be introduced at this point, since the complexity of natural communities may not be fully represented by existing toxicological experiments. As a matter of fact, benthic organisms may be locally significantly overexposed in regions where corrosion or low-order detonation leave exposed munition material on the seafloor. In addition, there seems to be growing evidence about the fact that munition-related chemicals can drive sublethal genetic and metabolic effects in aquatic organisms such as fishes.

For this reason, there exists a strong need to examine the occurrence of MC in natural ecosystems, in order to clarify potentially hidden effects; all this is hampered by the analytical difficulty in extracting and analyzing MC in tissues.

Once released from solid explosives, TNT is rather quickly transformed by bacteria and fungi to various amino-toluene derivatives, but most likely does not undergo a full mineralization [5]. Indeed, even though photochemical and abiotic reaction pathways may result in more complete degradations, the original aromatic structure can be preserved in reaction products such as nitrobenzenes.

From the analytical point of view, although the TNT derivate products can provide indications of 'dominant reaction pathways' (and therefore serve as an important tool for evaluating environmental fate and transport controls), their identification represents a real challenge.

Reaction rates of RDX and HMX, although tending to be slow, have greater chances to proceed towards the full mineralization with regard to TNT, ultimately resulting in the destruction of the original compounds [5]. There is also some evidence for mineralization and ring cleavage in natural samples and mesocosms [55, 56].

6.4.3 Geopolitical Aspects

The uncertainties in the specific nature of most MEC, the exact position of their disposal, and the unknown degree of corrosion and decomposition of explosive/chemical warfare agents make predicting the risk of detonation or the magnitude of possible releases of contaminants in the marine environment an extremely difficult task. In addition to this, the known presence on the seafloor of large amounts of military ordnance and warfare agents has produced a series of specific procedures and national regulations that may highly differ from country to country. As a result, when it comes to clearing operations, almost each nation and each Navy adopts different strategies and procedures, which are then combined in a geopolitical puzzle.

Nevertheless, some recent international projects have paved the way, devising a series of good practices for use when facing the problem of munitions dumped at sea, making it very clear that the problem needs to be accounted for using a holistic approach involving several different disciplines, an effort well beyond the specific reductionist one that has generally characterized this subject in the past.

As an example, during the CHEMSEA project, a broad series of interdisciplinary studies were conducted in the Baltic Sea area. These included current measurements, salinity and temperature distribution, bottom survey (side-scan sonar images), chemical analyses, biota communities' structure, and biomarker response of organisms to environmental stress. The project confirmed that MEC degradation products may be different in the environment compared to what is predicted by theoretical chemistry, as proved by Beldowski et al. [31] surveying more than $1500\,km^2$ of sea bottom and commencing 220 ROV sampling/identification missions.

Again in the Baltic region, the MODUM initiative, supported by NATO's Science for Peace and Security (SPS) Programme, established a cost-effective monitoring network [57] to observe munitions dumpsites using research vessels, AUVs, and ROVs. MODUM focused on technology application, rather than environmental studies, but nevertheless continued the studies done in CHEMSEA developing new portable analytical methods, reducing costs and time of degradation products analysis, and paving the way for new high-resolution numerical models for tracing the transport of contaminants from dumpsites through resuspension and currents (contaminated sediments from dumpsites can travel several kilometres in favourable meteorological forcing situations).

Funded by the EU, the recently launched project DAIMON will be capitalizing on past experiences, proposing a risk management tool so that decision-makers can evaluate the risks and benefits of various options in selected case sites. The tool will be based both on well-tested methods from the past and on novel, focused biological and chemical methods and risk assessment procedures to be developed within the project.

In the United States, the HUMMA project (*Hawai'i Underwater Military Munitions Assessment*, www.hummaproject.com) recently provided an extremely large integrated dataset, combining sonar, photographic, time-lapse acquisitions, and environmental data with the results of chemical analyses of physical specimens to investigate basic questions such as munitions casings integrity, distribution of chemicals in the sediments, and level of toxicity in animals, focusing on a large dumping area south of the island of Oahu, Hawai'i.

UXO and MEC in seas and oceans may be associated not only with deliberate dumping but also with other military activities, such as mining, sinking, or shipwrecks, making their management a matter of broad convert.

In the Pacific alone, there are about 4000 shipwrecks associated with WW II, some of them containing unknown amounts of ordnance and oil on board. In some cases, wrecks are heavily corroded and only 40% of the original hulls remain, which leaves the UXO exposed to seawater [58].

Hull plates were approximately 25 mm thick on ships from this era, and the corrosion rate (approximately 0.1 mm/year in seawater) may vary due to several factors, sometimes more than doubling due to physical and chemical factors (e.g. wave energy).

Moreover, there is no need for complete corrosion to release ammunition held inside wrecks, as partly corroded ships may either break under their own weight, or openings may be created due to internal stresses in the hull.

Further, baseline corrosion rates do not account for natural events, such as tectonic activity or cyclones, which ultimately caused the leaks in the case of *Mississinewa* [59], although studied only regarding fuel and oil leakages.

Possibly, the most known and clear example of danger to life, properties, and national infrastructures that may be brought by MEC within wrecks is provided by the case of the SS *Richard Montgomery*. The ship was wrecked in August 1994 in the Thames Estuary, carrying approximately 7000 tonnes of explosives, 3000 tonnes of which could not be removed before the ship broke its back, and are still on board in the forward section. The wreck remains on a 15-m contour, with the masts protruding out of the water, and significant corrosion and structural damage now probably presents a threat of triggering a mass detonation.

The SS *Richard Montgomery* lies between two busy shipping lanes leading to the major container facility at Thamesport and the London Gateway Port in the Thames. Moreover, the coastal towns of Sheerness and Southend-on-Sea lie to the south and north of the wreck. To further complicate the situation, the Isle of Grain Power Station, a liquefied natural gas terminal and a decommissioned oil refinery are all within the reach of any possible blast. The current wreck management strategy of non-interference requires periodic re-evaluations, to ensure the safety risk to the public is minimized. That policy is now being re-evaluated, thanks to the impact of time on the structure.

During the past few decades, however, a series of efforts have been undertaken to collect information about potentially hazardous wrecks. This includes the RUST (Resource and Under Sea Threats) initiative, which produced a database containing information on potentially hazardous wrecks in US waters [60]. Other countries such as Japan and France have also started similar initiatives [58]. According to RUST, wrecks built after 1910, steel hulled and with greater than 1000 gross tonnes, should be considered potentially dangerous in terms of releasing oil and cargo, including MEC [61].

Baltic countries have created a special working group on the environmental risks of hazardous submerged objects within HELCOM, named SUBMERGED. This is intended to evaluate the impact of wrecks, munitions, and other hazardous material on the sea bottom environment.

The rules for the classification of wrecks have been also coded by the Finnish Environment Institute (SYKE), on the basis of the Finnish register of wrecks which has existed since 1991 (containing more than 1000 objects) and of the Swedish register (containing 256 wrecks identified as hazardous to the marine environment and dumped in the region of the Kattegat and the Baltic Sea).

In Polish EEZ, about 1000 wrecks have been located, mostly originating from WW II (e.g. in Gdańsk Bay between 18 and 37 U-boats or remnants are estimated; see [62]). As part of the studies on the impact of selected wrecks on the Baltic Sea's ecosystem, which was carried out between 1998 and 2014, a total of 31 shipwrecks were examined (in 4 cases, contamination was found in the direct surroundings and on the wrecks).

All in all, the results outlined by these few examples confirm that MEC presence in the oceans represents a problem that is strongly case-specific and, very often, far beyond what a single country (or even a decently funded Project) can attempt to solve.

6.5 Moving Forward

6.5.1 Global Collaboration

Studies regarding dumped munitions are currently taking place in several areas, including the Baltic Sea, Hawaii areas, and the Adriatic Sea. However, there is little or no information about activities elsewhere, although munitions dumpsites are located in many areas of the world. Information exchange between existing projects is generally low, scattered, and usually limited to publication or presentation at conferences where this topic is not the main focus. It results in separated efforts, focused on small parts of the problem, while the obtained results are hard to extrapolate for other areas and remain site-specific. There is a need to create a consolidated group of scientists, practitioners, and decision-makers who can work together to tackle the problem under common frameworks.

At the moment, within the EU there seems to be a lack of specific funding schemes or calls aimed at addressing specifically the MEC problem. In the US context, the US Strategic Environmental Research and Development Program (SERDP) and the Environmental Security Technology Certification Program (ESTCP), both under the US Department of Defense (DoD) umbrella, funded a considerable number of projects, especially on detection, mapping, and modeling migration (https://www.serdp-estcp.org/Program-Areas/Munitions-Response/Munitions-Underwater/).

In parallel, within NATO, the Science and Technology Organization has been examining the environmental impact of munitions on the environment through a series of studies, symposia, and land-based demonstrations. These also demonstrated that, differently from land-based activities, munitions sea-dumping is still a problem that needs to be properly accounted for.

As a combined result, nowadays there does not seem to be enough attention on the MEC problems to allow a truly international partnership. Present opportunities for joint research are limited to a single country or small regions – like the Baltic countries or North Sea countries, for example. Joint studies, personnel exchanges, and visits are a must for the community to grow and to build capacity to really solve the problem. However, despite some recent efforts described here, they all suffer from scattered small funding and are not coordinated under a larger umbrella.

On the contrary, we believe that several specific challenges related to MEC are calling for a more 'global' approach, such as the following:

(i) MEC are very often distributed in marine areas that are pertinent to several nations or threaten international infrastructures, such as new oil and gas pipelines.
(ii) Sampling strategies and instrument specifics (sonars, ROVs, AUVs, magnetometers, etc.) should be the result of concerted actions.
(iii) Protocols to detect warfare agents and degradation products in the physical and biological environments should be agreed on at the international level, including standardization laboratory inter-comparison and standard reference benchmarks.

(iv) Multi-year monitoring programmes, allowing for the collection of long-time data series, necessary for thorough investigations, and identification of temporal trends, can only result from large and participative projects.
(v) Only major shared efforts will allow an advance in the application of high-resolution *ad hoc* numerical modeling tools, capable of producing scenarios for interventions and supporting risk assessment procedures, as well as producing modeling dealing with the chemical speciation of the conventional and chemical warfare agents at sea.
(vi) New findings and technology need to be tested and inter-compared during sea trials that require joint efforts, more likely to be carried out in an international context of leveraging.

This is even more true if we refer to the pillars on which the EU *Blue Growth* initiative (ec.europa.eu/maritimeaffairs/policy/blue_growth_en) is standing. All sectors considered as essential in the *Blue Growth* strategy (aquaculture, coastal tourism, marine biotechnology, ocean energy, seabed mining) are clearly cutting through the MEC problem. All the components needed to provide knowledge, legal certainty, and security to the so-called Blue Economy (marine knowledge, Maritime Spatial Planning (MSP), integrated marine surveillance) are clearly also necessary for a thorough assessment of MEC.

6.5.2 Recent Global EU and NATO Efforts

In an effort to set the MEC issue in a truly international framework, in November 2015 the European *Joint Programming Initiative – Healthy and Productive Seas and Oceans* (JPI Oceans, a high-level coordinating and integrating strategic platform, open to all EU Member States and Associated Countries) launched a joint action named *Munitions in the Sea* (http://www.jpi-oceans.eu/munitions-sea), which will provide knowledge-based support to operators and policy makers. The aim of the action involving several EU countries is to assess risks and describe case studies, define priorities, and suggest common intervention options. The idea of such a coordinated transnational effort aims at increasing the efficiency and effectiveness of interventions to emergencies, often adopted at national or local levels, by sharing experiences and skills across Europe. Outcomes will be used to support decision-makers in the identification, monitoring, and elimination of threats through more systematic and shared approaches across Europe.

As mentioned, NATO has become increasingly aware of the issues arising from the problems of sea-dumped munitions thanks to the series of studies and symposia held under the auspices of the NATO Science and Technology Organization. NATO had identified this area as one needing attention and held a workshop in Bulgaria on the Black Sea Coast at Varna to review the status and discuss approaches to the areas already mentioned. Several groups attended, building links between those listed and the traditional NATO science and technology community. It is to be hoped that this may be maintained and developed through additional joint action, since the JPI-O Group took an active part in this NATO activity.

Concurrently, during the spin-up phase, the action *Munitions in the Sea* has been involving in its activities the Navies of different countries and has also been establishing contacts with the US experts in the field (e.g. the University of Hawai'i Manoa).

6.5.3 Advantages of Joint Efforts

When dealing with the MEC problem, it has always been very difficult to produce a series of effective shared protocols and recommendations. This is due both to the scientific nature of the problem, which *per se* incorporates several disciplines ranging from hydrodynamics, biology, chemistry, marine technology, etc., and to the fact that the topic is often treated as a national problem by each country and its Navy. On the contrary, there are several outcomes that the umbrella of a concerted action, favouring joint efforts between different countries and Navies, could bring along.

Scientific advances. Greater and international scientific support to MEC-related issues would contribute to the increase of the knowledge about specific subjects that presently represents a challenge (e.g. understanding and modeling the chemical speciation of chemical agents at sea, assessing the possible distribution of contaminants using high-resolution numerical models, etc.). Joint collaborations would lead to safer planning and removal of munitions from the sea floor, and will also contribute to establish more shared working scenarios, providing risk-assessment estimates and eventually an internationally agreed approach.

Technology innovation. An internationally shared approach will establish more easily accepted and standardized procedures relevant to the military and commercial end users, which may turn out to be more directly applicable in the context of Maritime Spatial Planning (MSP) activities. Examples may be agreed protocols of interventions for munitions at sea (starting from an interactive map of a working area and the type of dumped munitions, addressing the main risks associated in case of intervention); shared protocols to determine toxicity (especially chronic) on marine organisms (e.g. the development of specific biomarkers), including the development of portable methods; the definition of maps of risks integrating the *in situ* available information with numerical models in selected areas; the definition of a knowledge-based support for interventions in terms of options, guidelines, scenarios, and state-of-the-art methodology that can be adopted by the different Navies.

Research infrastructures and data. A wider, international action in the field of the munitions in the sea will promote and stimulate initiatives to optimize interventions, sharing already existing infrastructures (supercomputing centers, shiptime, AUVs, ROVs, etc.) and favouring open access to existing data.

Human capacities. A global approach to the problem of sea-disposed munitions would generate a higher interaction among different levels of governance, sectors, and disciplines. More effective specific actions would thus be oriented towards off-shore operators and public authorities' training, fishermen training, public awareness of sea-dumped MEC, disseminating the message within the education system, etc.

The above-mentioned areas of cooperation are still under development. At the moment, several initiatives are ongoing to bring together communities working on MEC in different areas of the world. It includes ongoing national and EU projects, JPI Oceans group, both HELCOM and OSPAR Commissions, US projects. Relevant stakeholders have been asked for collaboration, by several communities – i.e. Baltic region, the United States, and the Mediterranean Sea – while a collection of end users' priorities at national level in EU countries and the United States is now been obtained.

This will be used to outline the priorities that will guide the next steps of the action towards the definition of an implementation plan.

However, it will be the responsibility of the international actors, especially governmental bodies, funding agencies, and intergovernmental organizations, to set up proposals for activities, or address such activities via specific calls, involving appropriate partners (including non-governmental ones, e.g. IDUM, www.underwatermunitions.org) and helping to build links amongst the various groups and participants. This is the only possible path towards a truly effective action on MEC that can join strengths for mutual benefits.

Glossary

AUV autonomous underwater vehicle. A robot that can travel underwater without requiring input from an operator
CHEMSEA *chemical Munitions Search and Assessment,* http://www.chemsea.eu/
CWA chemical warfare agent
DAIMON *decision Aid for Marine Munitions,* https://www.daimonproject.com/
DMM disposed (or discarded) military munitions. Munitions that were not used and therefore dumped
EMI electromagnetic induction systems, a type of electromagnetic sensor employed for MEC detection
GLONASS Russian Global Navigation Satellite System. A space-based satellite radio navigation system, alternative to GPS
GNSS global Navigation Satellite System (e.g. GPS)
GPS Global Positioning System. A satellite-based radio navigation system, providing geolocation and time anywhere on earth
HUMMA *Hawai'i Undersea Military Munitions Assessment,* http://www.hummaproject.com/
HELCOM *Baltic Marine Environmental Protection Commission,* www.helcom.fi
INS Inertial Navigation System. A navigation aid using a computer, rotation, and motion and magnetic sensors to continuously calculate by dead reckoning position, orientation, and velocity of a moving object without the need for external references
JPI Oceans the Joint Programming Initiative Healthy and Productive Seas and Oceans. An intergovernmental initiative to enable cooperation in marine and maritime research

LBL long base line, an acoustic systems used to determine position underwater
LLS laser line scanners, optical sensor used to image the seafloor
MC Munitions compounds or constituents
MEC Munitions and explosives of concern. It includes both UXO and DMMs
OSPAR The Convention for the Protection of the Marine Environment of the North-East Atlantic
POP Persistent organic pollutants
ROV remotely operated vehicle. It is a tethered, unoccupied underwater mobile device
UXO Unexploded Ordnance. Munitions that were fused and fired, as a result of firing ranges or military action

Acknowledgements

The authors are indebted to Dr. Aaron Beck (GEMOAR Germany) for the careful reading of the proof and useful suggestions on several parts of the manuscript. Part of this text was developed in the frame of the BSR project DAIMON, funded by the European Union (European Regional Development Fund) under the Interreg Baltic Sea Region Programme 2014–2020, project #R013 DAIMON, co-funded by the Ministry of Science and Higher Education (Poland) from the 2016–2019 science funding No. 3623/INTERREG BSR/2016/2 allocated for the implementation of international co-financed project DAIMON. SC thanks the coordinator of the JPI Oceans initiative 'Munitions in the Sea', Dr. E. Campana, Dr. P.F. Moretti (CNR), Dr. E. Amato (ISPRA), and Commander G. Modugno (Italian Navy) for the fruitful discussions. Besides, SC thanks the National Research Council for the support provided by the Short-Term Mobility program (2018), that allowed him to spend a period at the Applied Research Laboratory, University of Hawaii (the United States) in early 2018.

References

1 Baine, E., Simmons, M. (2005). Mitigating the possible damaging effects of twentieth-centuryocean dumping of chemical munitions. In: 8th International Chemical Weapons Demilitarization Conference, CWD 2005, Edinburgh, Scotland, United Kingdom (12–14 April).
2 Carton, G. and Jagusiewicz, A. (2011). Historic disposal of munitions in U.S. and European coastal waters, how historic information can be used in characterizing and managing risk. *Mar. Technol. Soc. J.* 43: 16–32. https://doi.org/10.4031/MTSJ.43.4.1.
3 Amato, E., Alcaro, L., Corsi, I. et al. (2006). *An Integrated Ecotoxicological Approach to Assess the Effects of Pollutants Released by Unexploded Chemical Ordnance Dumped in the Southern Adriatic (Mediterranean Sea)*. Heidelberg: Springer.
4 Carniel, S., Beldowski, J., and Cumming, A. (2016). Munitions in the sea: time for a global action. *Seal. Technol.* 58 (1): 37–39.
5 Beck, A.J., Gledhill, M., Schlosser, C. et al. (2018). Spread, behavior, and ecosystem consequences of conventional munitions compounds in coastal marine waters. *Front. Mar. Sci.* 5: 141. https://doi.org/10.3389/fmars.2018.00141.

6 Böttcher, C., Knobloch, T., Rühl, N.-P. et al. (2011). Munitionsbelastung der Deutschen Meeresgewässer – Bestandsaufnahme und Empfehlungen (Stand 2011). Meeresumwelt Aktuell Nord- und Ostsee, 2011/3. Hamburg; Rostock: Bundesamt für Seeschifffahrt und Hydrographie (BSH).

7 Arison, L.H. (2013). *European Disposal Operations: The Sea Disposal of Chemical Weapons.* CreateSpace Independent Publishing Platform.

8 Knobloch T., Bełdowski J., Böttcher C. et al. (2013). Chemical Munitions Dumped in the Baltic Sea. Rep. of the *ad hoc* Expert Group to Update and Review the Existing Information on Dumped Chemical Munitions in the Baltic Sea (HELCOM MUNI) Baltic Sea Environmental Proceedings.

9 "FiscalYear (2009). Defense Environmental Programs Annual Report to Congress," Preparedby the Office of the Under Secretary of Defense for Acquisition, Technology andLogistics, Washington, DC, April 2010, p. 85–120.

10 Duursma, E.K. and Surikov, B.T. (1999). *Dumped Chemical Weapons in the Sea, Options: A Synopsis.* Amsterdam, Netherlands: A.H. Heineken Foundation for the Environment.

11 Korendovych V. (2012). Enhancing abilities to counter WMD proliferation – Ukrainian experience. Paper presented at the Countering WMD Threats in Maritime Environment, Riga.

12 Beldowski, J., Been, R., and Turmus, E. (2017). *Towards the Monitoring of Dumped Munitions Threat (MODUM)*, NATO Science for Peace and Security Series, Series C: Environmental Security. Dordrecht, The Netherlands: Springer.

13 Czub, M., Kotwicki, L., Lang, T. et al. (2018). Deep sea habitats in the chemical warfare dumping areas of the Baltic Sea. *Sci. Total Environ.* 616–617: 1485–1497.

14 Missiaen, T., Soderstrom, M., Popescu, I., and Vanninen, P. (2010). Evaluation of a chemical munition dumpsite in the Baltic Sea based on geophysical and chemical investigations. *Sci. Total Environ.* 408 (17): 3536–3553. doi: 10.1016/j.scitotenv.2010.04.056.

15 Sanderson, H., Fauser, P., Thomsen, M., and Sorensen, P.B. (2008). Screening level fish community risk assessment of chemical warfare agents in the Baltic Sea. *J. Hazard. Mater.* 154 (1-3): 846–857. https://doi.org/10.1016/j.jhazmat.2007.10.117.

16 Missiaen, T., Henriet, J.P., and Team, P.P. (2002). Chemical munitions off the Belgian coast: an evaluation study. In: *Chemical Munition Dump Sites in Coastal Environments* (ed. T. Missiaen and J.P. Henriet). Federal Office for Scientific, Technical and Cultural Affairs (OSTC).

17 Edwards, M. and Carniel, S. (2017). Addressing munitions in the sea, lessons from the Hawaii undersea military munitions assessment. *Seal. Technol.* 58 (6): 29.

18 Edwards, M.H., Wilkens, R.H., Kelley, C., and Carton, G. (2012). Methodologies for surveying and assessing deep water munitions disposal sites. *Mar. Technol. Soc. J.* 46: 51. https://doi.org/10.4031/MTSJ.46.1.6.

19 Andrulewicz, E. (1996). War gases and ammunition in the polish economic zone of the Balic Sea. In: *Sea-Dumped Chemical Weapons: Aspects, Problems and Solutions*, vol. 7 (ed. A.V. Kafka), 9–15. Kluwer Academic Publishers.

20 George, R., Wild, B., Li, S. et al. (2015). *Recovery Corrosion Analysis, and Characteristics of Military Munitions from Ordnance Reef (HI-06)*. Report, 61.
21 Jurczak, W. and Fabisiak, J. (2017). Corrosion of ammunition dumped in the Baltic Sea. *J. KONBiN* 41: 227–246. https://doi.org/10.1515/jok-2017-0012.
22 Overfield, M.L. and Symons, L.C. (2009). The use of the RUST database to inventory, monitor, and assess risk from undersea threats. *Mar. Technol. Soc. J.* 43: 33–40. https://doi.org/10.4031/MTSJ.43.4.9.
23 Porter, J.W., Barton, J.V., and Torres, C. (2011). Ecological, radiological, and toxicological effects of naval bombardment on the coral reefs of Isla de Vieques, Puerto Rico. *Warfare Ecol.* 65–122. https://doi.org/10.1007/978-94-007-1214-0_8.
24 Edwards, M.H., Shjegstad, S.M., Wilkens, R. et al. (2016). The Hawaii Undersea Military Munitions Assessment. *Deep. Res. Part II Top. Stud. Oceanogr.* 128: 4–13. https://doi.org/10.1016/j.dsr2.2016.04.011.
25 Francken, F. and Hafez, A. (2009b). A case study in modeling dispersion of yperite and CLARK I and II from munitions at Paardenmarkt, Belgium. *Mar. Technol. Soc. J.* 43: 52–61.
26 Zaborska, A., Winogradow, A., and Pempkowiak, J. (2014). Caesium-137 distribution, inventories and accumulation history in the Baltic Sea sediments. *J. Environ. Radioact.* 127: 11–25.
27 Pempkowiak, J., Beldowski, J., Pazdro, K. et al. (2002). The contribution of the fine sediment fraction to the fluffy layer suspended matter (FLSM). *Oceanologia* 44 (4): 513–527.
28 Pennington, J.C. and Brannon, J.M. (2002). Environmental fate of explosives. *Thermochim. Acta* 384: 163–172.
29 Juhasz, A.L. and Naidu, R. (2007). Explosives: fate, dynamics, and ecological impact in terrestrial and marine environments. *Rev. Environ. Contam. Toxicol.* 191: 163–215.
30 Sæbø, T.O., Hansen, R.E., and Lorentzen, O.J. (2015). Using an interferometric synthetic aperture sonar to inspect the Skagerrak World War II chemical munitions dump site. In: *OCEANS 2015 – MTS/IEEE Washington*, 1–10. Washington, DC.
31 Beldowski, J., Klusek, Z., Szubska, M. et al. (2016). Chemical munitions search and assessment-an evaluation of the dumped munitions problem in the Baltic Sea. *Deep-Sea Res. II* 128: 85–95. https://doi.org/10.1016/j.dsr2.2015.01.017.
32 Kotwicki, L., Grzelak, K., and Beldowski, J. (2016). Benthic communities in chemical munitions dumping site areas within the Baltic deeps with special focus on nematodes. *Deep-Sea Res. II* 128: 123–130. https://doi.org/10.1016/j.dsr2.2015.12.012.
33 Söderström, M., Östin, A., Qvarnström, J. et al. (2018). Chemical analysis of dumped chemical warfare agents during the MODUM project. In: *Towards the Monitoring of Dumped Munitions Threat (MODUM): A Study of Chemical Munitions Dumpsites in the Baltic Sea* (ed. J. Bełdowski, R. Been and E.K. Turmus), 71–103. Dordrecht: Springer Netherlands.
34 Belden, J.B., Ownby, D.R., Lotufo, G.R., and Lydy, M.J. (2005). Accumulation of trinitrotoluene (TNT) in aquatic organisms: Part 2 – Bioconcentration in aquatic invertebrates and potential for trophic transfer to channel catfish (Ictalurus punctatus). *Chemosphere* 58: 1161–1168. https://doi.org/10.1016/j.chemosphere.2004.09.058.

35 Lotufo, G.R. and Lydy, M.J. (2005). Comparative toxicokinetics of explosive compounds in sheepshead minnows. *Arch. Environ. Contam. Toxicol.* 49: 206–214. https://doi.org/10.1007/s00244-004-0197-7.

36 Ownby, D.R., Belden, J.B., Lotufo, G.R., and Lydy, M.J. (2005). Accumulation of trinitrotoluene (TNT) in aquatic organisms: part 1 – Bioconcentration and distribution in channel catfish (Ictalurus punctatus). *Chemosphere* 58: 1153–1159. https://doi.org/10.1016/j.chemosphere.2004.09.059.

37 Rosen, G. and Lotufo, G.R. (2007). Toxicity of explosive compounds to the marine mussel, Mytilus galloprovincialis, in aqueous exposures. *Ecotoxicol. Environ. Saf.* 68: 228–236. https://doi.org/10.1016/j.ecoenv.2007.03.006.

38 US, E.P.A. (2012). EPA Federal Facilities Forum Issue Paper: Site Characterization For Munitions Constituents. In: *EPA-505-S-11-001*.

39 Oyler, K.D., Meta, N., and Chen, G. (2015). *Overview of Explosive Initiators*, 24. Picatinny Arsenal, NJ: Army Armament Research Development and Engineering Center https://doi.org/10.21236/ADA625185.

40 Yamamoto, H., Morley, M.C., Speitel, G.E., and Clausen, J. (2004). Fate and transport of high explosives in a sandy soil: adsorption and desorption. *Soil Sediment Contam.* 13: 459–477.

41 McGrath C.J. (1995). Review of Formulations for Processes Affecting the Subsurface Transport of Explosives. *Technical Rep. IRRP-95-2*. U.S. Army Corps of Engineers, Waterways Experiment Station, Vicksburg, MS.

42 Gregory, K.B., Larese-Casanova, P., Parkin, G.F., and Scherer, M.M. (2004). Abiotic transformation of hexahydro-1,3,5-trinitro-1,3,5-triazine by fell bound to magnetite. *Environ. Sci. Technol.* 38: 1408–1414.

43 Wang, P.-F., George, R. D., Wild, W., and Liao, Q. (2013). Defining Munition Constituent (MC) Source Terms in Aquatic Environments on DoD Ranges (ER-1453). SSC Pacific Technical Report 1999. 130 pp.

44 Paka, V. and Zhurbas, V. (2008). Estimates of eddy diffusivity in bottom boundary layer of the Bornholm deep. In: *2008 IEEE/OES US/EU-Baltic International Symposium*, 405–407. IEEE.

45 Ro, K.S., Venugopal, A., Adrian, D.D. et al. (1996). Solubility of 2,4,6-Trinitrotoluene (TNT) in Water. *J. Chem. Eng. Data* 41: 758–761. https://doi.org/10.1021/je950322w.

46 Lynch, J.C., Myers, K.F., Brannon, J.M., and Delfino, J.J. (2001). Effects of pH and temperature on the aqueous solubility and dissolution rate of 2,4,6-trinitrotoluene (TNT), Hexahydro-1,3,5-trinitro-1,3,5-triazine (RDX), and Octahydro-1,3,5,7-tetranitro-1,3,5,7-tetrazocine (HMX). *J. Chem. Eng. Data* 46: 1549–1555. https://doi.org/10.1021/je0101496.

47 Pichtel, J. (2012). Distribution and fate of military explosives and propellants in soil: a review. *Appl. Environ. Soil Sci.* 2012: 1–33. https://doi.org/10.1155/2012/617236.

48 Bausinger, T., Bonnaire, E., and Preuß, J. (2007). Exposure assessment of a burning ground for chemical ammunition on the GreatWar battlefields of Verdun. *Sci. Total Environ.* 382: 259–271. https://doi.org/10.1016/j.scitotenv.2007.04.029.

49 Tornero, V. and Hanke, G. (2016). Chemical contaminants entering the marine environment from sea-based sources: a review with a focus on

European seas. *Mar. Pollut. Bull.* 112: 17–38. https://doi.org/10.1016/j.marpolbul.2016.06.091.

50 Callaway, A., Quinn, R., Brown, C.J. et al. (2011). Trace metal contamination of Beaufort's Dyke, North Channel, Irish Sea: a legacy of ordnance disposal. *Mar. Pollut. Bull.* 62: 2345–2355. https://doi.org/10.1016/j.marpolbul.2011.08.038.

51 Gebka, K., Bełdowski, J., and Bełdowska, M. (2016). The impact of military activities on the concentration of mercury in soils of military training grounds and marine sediments. *Environ. Sci. Pollut. Res.* 23: 23103–23113. https://doi.org/10.1007/s11356-016-7436-0.

52 Green, A., Moore, D., and Farrar, D. (1999). Chronic toxicity of 2, 4, 6-trinitrotoluene to a marine polychaete and an estuarine amphipod. *Environ. Toxicol. Chem.* 18: 1783–1790.

53 Lotufo, G.R., Farrar, J.D., Innouye, L.S. et al. (2001). Toxicity of sediment-associated nitroaromatic and cyclonitramine compounds to benthic invertebrates. *Environ. Toxicol. Chem.* 20: 1762–1771.

54 Nipper, M., Carr, R.S., Biedenbach, J.M. et al. (2001). Development of marine toxicity data for ordnance compounds. *Arch. Environ. Contam. Toxicol.* 41: 308–319.

55 Montgomery, M.T., Boyd, T.J., Smith, J.P. et al. (2011). 2,4,6-Trinitrotoluene mineralization and incorporation by natural bacterial assemblages in coastal ecosystems. In: *Environmental Chemistry of Explosives and Propellant Compounds in Soils and Marine Systems: Distributed Source Characterization and Remedial Technologies*, ACS symposium series (ed. M.A. Chappell, C.L. Price and R.D. George), 171–184. Washington, DC: American Chemical Society.

56 Smith, R.W., Tobias, C., Vlahos, P. et al. (2015). Mineralization of RDX-derived nitrogen to N_2 via denitrification in coastal marine sediments. *Environ. Sci. Technol.* 49: 2180–2187. https://doi.org/10.1021/es505074v.

57 Lang, T., Kotwicki, L., Czub, M. et al. (2018). The health status of fish and Benthos communities in chemical munitions dumpsites in the Baltic Sea. In: *Towards the Monitoring of Dumped Munitions Threat (MODUM): A Study of Chemical Munitions Dumpsites in the Baltic Sea* (ed. J. Bełdowski, R. Been and E.K. Turmus), 129–152. Dordrecht: Springer Netherlands.

58 Barrett, M.J. (2011). Potentially polluting shipwrecks, spatial tools and analysis of WWII shipwrecks. Masters Project, Duke University.

59 U.S. Navy (2004). U.S. NAVY Salvage Report USS Mississinewa Oil Removal Operations. Naval Sea Systems Command.

60 Basta, D. (2010). Opening remarks. International Corrosion Workshop, Newport News, VA.

61 Symons L. (2010). Identification of data gaps and development of research priorities. International Corrosion Workshop, Newport News, VA.

62 Komorowski, A. (ed.) (2003). *Historical treasures of the Gulf of Gdansk, Torun*, 126.

7

Environmental Assessment of Military Systems with the Life-Cycle Assessment Methodology

Carlos Ferreira, Fausto Freire, and José Ribeiro

ADAI-LAETA, Department of Mechanical Engineering, University of Coimbra, Portugal

Environmental and toxicological consequences associated with ammunition have been divorced from the concerns of the general population and authorities due to their intended application in war or conflict scenarios. However, the perception and the relevance of the problem can be changed completely when it is considered that only a small percentage of ammunitions are used in war scenarios. In reality, the majority of munitions are demilitarized or used in live-fire training, contributing significantly to soil, water, and air contamination with (unburnt or partially burnt) energetic materials, heavy metals, and other contaminants. Munitions used in battle have similar effects, but these are generally less common although perhaps more complex.

Research on munitions is usually related to safety (e.g. insensitive munitions) and performance improvement (e.g. high energy density and high detonation velocity energetic materials, etc.), involving different but complementary fields of research. However, due to the increase in environmental awareness, the emissions produced by unreacted, or partially reacted, energetic materials associated with ammunition use, the heavy metals deposited, and the amount of those contaminants in the soil and water have started to be measured [1–10]. Some studies have also researched the amount of lead intake by birds [11, 12] and the human exposure to lead by direct inhalation at small-calibre ammunition shooting ranges [13].

The aforesaid studies, which are based on monitoring individual aspects, lack the holistic view that it is possible to address within a life-cycle perspective. Certainly the evaluation of heavy metal toxicity (e.g. lead) to some species helps understand the importance and increase the awareness associated with ammunition contamination, although that can be insufficient to comprehend the potential environmental burdens as other substances pose a significant influence on different environmental impacts. To completely comprehend the consequences related to the use of ammunition, it is necessary to undertake the quantification of potential impacts on human health and ecosystems. Therefore, it is important to use the data from the characterization of those emissions to

Energetic Materials and Munitions: Life Cycle Management, Environmental Impact and Demilitarization, First Edition. Edited by Adam S. Cumming and Mark S. Johnson.
© 2019 Wiley-VCH Verlag GmbH & Co. KGaA. Published 2019 by Wiley-VCH Verlag GmbH & Co. KGaA.

quantify the potential burdens. In order to assess the environmental impacts of ammunition, it is necessary to address all emissions of heavy metals, energetic materials, and other substances, and to consider other factors beyond the impacts on human health and different individual species that exist in ecosystems. These must include other potential environmental impacts (e.g. global warming, acidification).

It is important to note that other life-cycle phases of military activities also have associated environmental impacts that should not be neglected. Sources of contamination from ammunition production are usually associated with explosives manufacture processes (synthesis, casting, curing, and machining), inadequate storage practices, and inadequate disposal of contaminated wastewaters [14]. With respect to ammunition decommission, the main environmental burdens are reported to come from open burning/open detonation. Substantial residues are dispersed, and these are associated mostly with low-order detonations or ammunitions thrown out from the burn pits (resulting in unexploded ordnance, UXO) [14]. Decommission of ammunition by incineration can also present some environmental concerns, mainly due to indirect emissions resulting from the consumption of energy to clean the flue gases from the energetic materials combustion or thermal treatment [15].

The main problem affecting the pursuit of environmental assessment of military activities is the selection of a suitable methodology to assess and, if possible, mitigate those impacts in a life-cycle perspective. The life-cycle assessment (LCA) methodology can be an appropriate tool to help in quantifying the environmental and toxicological impacts associated with the whole life cycle of military activities or products. Therefore, this chapter intends to shed some light on this methodology in general and for the particular case of military systems.

This chapter is structured in five parts, including this introductory section. The second part presents an overview of the LCA methodology which describes its main purpose, the standards and guidelines that cover it, and the definition of the life-cycle phases. The third part describes the life-cycle phases in detail with the assistance of a case study involving ammunition. The case study in which the LCA methodology was used enables us to understand the advantages of using this tool to ascertain the environmental burden of military systems, in particular by demonstrating the importance of using a life-cycle approach to avoid unexpected trade-offs or adverse consequences. In the fourth part is presented the limitations of the LCA methodology, and the last part draws some conclusions.

7.1 Overview of the Life-Cycle Assessment Methodology

7.1.1 Life-Cycle Thinking

In the beginning of the 1990s, a shift in the focus of environmental concern in industry was observed. Environmental strategy for industry began to introduce a way of thinking that would go beyond end-of-process approaches and emission control. Therefore, industry started to include a path to the environmental optimization of products rather than only focusing on the minimization of emissions from industrial processes. This was due to the environmental concerns being verified for certain products during use and as waste [16].

This change in perception helped establish the concept of life-cycle thinking, which inspired the holistic perspective considered in LCA. The essence of life-cycle thinking is that processes and products do not exist in isolation, but carry baggage and implications. Therefore, the focus should be not only on the process or product under study but also on the activities that are related to these processes, such as preceding activities (e.g. design, extraction, processing) or downstream activities (e.g. dismantlement, materials recovery, landfill).

The main repercussion produced by the life-cycle thinking is an awareness of the need to consider all life-cycle phases, and the inherent environmental burdens, of a product or system. The entire life cycle of a product, from raw material extraction and acquisition, through production of energy and material, to usage and end-of-life treatment and final disposal is considered (called cradle-to-grave analysis). The main aim of employing life-cycle thinking is to reduce the use of resources, identify potential improvements to products or activities across all life-cycle phases, and avoid shifting of burdens [17]. This thinking is a very broad framework in which environmental impacts and implications are treated only in a qualitative way or using simple scoring systems [18]. This is often adopted in the environmental policy of companies and governments [16].

Life-cycle thinking can be put into practice for policy making (e.g. green procurement, ecolabel, circular economy), and for business (e.g. marketing, ecodesign, carbon reduction) [18], with the use of tools that incorporate this way of thinking such as material flow analysis, input–output analysis, life-cycle costing, multicriteria analysis, and LCA.

An important advantage of using a life-cycle perspective is that it is possible to identify and possibly avoid the shifting of potential environmental burdens between life-cycle stages or individual processes, thus ensuring that all potential burdens are taken into account. The shifting of impacts can be associated with the introduction of a new technology: for instance, the introduction of electric vehicles can decrease direct emissions (exhaust emissions) but can increase the impact related to indirect emissions (those that are related to the production of electricity). Another example can be associated with the selection of a substance that presents a low toxicological impact during its use, but has a high impact associated with its production (when compared with conventional substances).

7.1.2 Life-Cycle Assessment

LCA methodology follows the life-cycle perspective in which the potential environmental impacts of a product system throughout its life cycle is assessed [19]. Figure 7.1 presents a simplified representation of the life-cycle phases of a product (or system) which LCA methodology handles as a chain of subsystems. These exchange inputs and outputs include extraction of materials (resources), production, use, end-of-life treatment, recycling, and final disposal (cradle-to-grave analysis) [19, 20].

Figure 7.1 Representation of the life-cycle phases of a product.

LCA methodology principles and requirements are largely covered principally in the ISO standards 14040 [19] – (Principles and Framework) and 14044 [20] – (Requirements and Guidelines). These standards were developed on the basis of international agreements, but they do not describe or specify the methodology in detail, for they are not intended for contractual or regulatory purposes, or for registration and certification [17]. Therefore, these standards are an overview of the principal elements of the LCA methodology that includes the outline and description of the four phases of the methodology, and the elements that need to be considered for each phase [17].

Figure 7.2 illustrates the four interrelated phases of LCA according to the ISO standards: goal and scope definition, life-cycle inventory (LCI), life-cycle impact assessment (LCIA), and interpretation. The first phase includes the definition of the goal and scope of the study, including the selection of the system boundaries and a functional unit. The depth and the extent of an LCA study can differ considerably depending on the goal defined. In the LCI analysis, the inputs and outputs necessary to meet the goals of the defined study are collected and compiled. In the LCIA, the inventory data is characterized into specific environmental impact categories according to selected LCIA methods. The purpose of LCIA is to provide additional information to help assess a product's environmental burden. It is important to mention that different LCIA methods will lead to different results (values, impact categories, and units) [21]. Interpretation is the final phase of the LCA methodology, in which the results are summarized and discussed as a basis for conclusions, recommendations, and decision-making in accordance with the goal and scope definition phase. The four life-cycle phases of the LCA methodology are presented in more detail in Section 7.3 with the help of case studies for military activities.

7.1.3 Purpose of Life-Cycle Assessment Studies

LCA studies can be applied to various aspects that can extend from product development to marketing issues. The main purpose of an LCA study is to comprehend which are the contributors to the environmental impacts, and identifying opportunities to improve the environmental performance of products. In the development phase of products, it is possible to apply an LCA since the conceptual phase (e.g. ecodesign), in order to consider which appropriate materials can be used in an earlier stage with the aim of achieving sustainable development [16]. For instance, the Ecodesign Directive (Directive 2009/125/EC) ensures that manufacturers consider the environmental burdens, including energy consumption along its lifetime, during the design phase of energy-related products, and LCA can be used to implement the measures needed for the ecodesign of these products [17]. Product design with improved environmental design choices of materials and technologies can also help with green procurement by helping to select those suppliers with an enhanced environmental performance [16]. At this stage, it is also possible to plan products based on a 'design for disposal', which means that is possible to design products taking into consideration their end-of-life in order to enable its reuse or recycling.

In general, the most noticeable application of LCA is the pursuit of improvement or optimization possibilities, principally in studies done for industry. These studies can be performed for existing products or for processes under development with the aim of asserting which are the contributors to environmental impact, identify which life-cycle phase presents the higher contributions to the impacts, and prevent shifting of impacts; comparing different products or technologies with the same function; and informing decision-makers in industry, government, or non-government organizations.

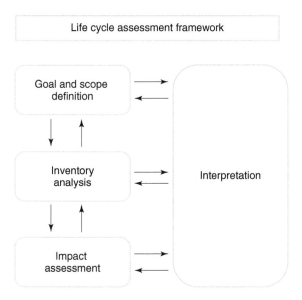

Figure 7.2 The four phases of the life-cycle assessment methodology. *Source:* Adapted from ISO 14040 [19].

Another application of LCA is for market communication through the implementation of eco-labelling and environmental product declaration programmes that offer a standardized message which is intended to increase the credibility of that information [16]. Green marketing can benefit consumers by enabling the purchase of environmentally aware products, while also benefiting the companies by providing a vehicle to promote a product with additional information. It can also support environmental protection by presenting an instrument that drives product development in a sustainable approach [16].

In the specific case of military activities, the LCA studies can allow decision-makers, whether shooting range managers, ammunition procurement officers, ammunition producers, or others, to become more aware of the main environmental and toxicological problems associated with the production, use, and disposal of ammunition. These studies help in defining strategies to manage or mitigate ammunition burdens and carry out tailored modifications to decrease the environmental impact. For instance, the application of LCA to ammunition enabled the practitioner to answer questions such as: What is contributing to the environmental and toxicological impact? Which life-cycle phase produces higher impact? Are new 'green' munitions producing lower environmental impact than traditional ones? Does the use of alternative materials raise new environmental concerns? Are the impacts significant and why? What further steps should be taken to improve ammunitions from a toxicological and environmental point of view? What technology shows lower environmental and toxicological impact in the disposal of ammunition?

7.2 The Four Phases of the LCA Methodology Applied to a Case Study

This section presents in detail the four phases for conducting an LCA study. The four phases are applied to a case study to help comprehend, in practice, how to perform an LCA study, which is presented in italic to help readers easily find the parts related to the case study. *The case study that was described in Ferreira et al. [22] is related to an ecodesign study for small-calibre ammunition (9 mm) whose principal objective was to identify the potential benefits of removing lead from the propellant and projectile of that ammunition. For that purpose, the LCA is applied to four types of small-calibre ammunition, which are combinations of two different projectiles (steel jacket and lead core; copper and nylon composite) and two types of primers (lead primer; non-lead primer).*

7.2.1 Goal and Scope

In the goal and scope phase, the main objective of an LCA study is defined, which include the clarification of the intended use and purpose of the study, as well as the audience interested in that study [16, 19]. The purpose of LCA studies for military systems can be formulated as general, or more specific, questions: What is contributing to the environmental and toxicological impact? Which life-cycle phase represents higher impact? Are new materials offering lower environmental impact than conventional ones? What further steps should be taken to enhance the environmental performance of ammunition? What technology shows lower environmental and toxicological impact for the disposal of ammunition?

The main objective of this case study is to assess and compare the environmental and toxicological impacts associated with the life cycle of four types of 9-mm ammunition. In more detail, the objectives are to (i) ascertain the most relevant life-cycle phases, (ii) identify the main contributors to the environmental and toxicological impact associated with four ammunitions types, and (iii) identify opportunities to improve the environmental performance of those munitions. Different decision-makers and procurement officers can have their specific interests and priorities addressed. The results delivered by the assessment for the small-calibre ammunition (or other type of ammunition) can be of interest for producers, range managers, government bodies, and other agencies related to defence. This will assist in defining strategies or policies to manage or mitigate ammunition burdens and carry out tailored modifications to decrease the impacts associated with military products and activities.

Based on the goal determined, the scope of the study is also defined. This includes the specification of the methodological assumptions that are necessary to perform the study, namely, the definition of functional unit, system boundaries, selection of impact categories, allocation procedures, amongst other requirements [16]. The definition of the goal and scope is a crucial part of the LCA studies as different methodological choices can provide different answers. However, the assumptions made in an LCA study should be transparent and not arbitrary [16]. The following subsections describe in general the functional unit and system boundaries, and indicate the assumptions made for the case study.

7.2.1.1 Functional Unit

The definition of the functional unit is of major importance in LCA studies as it is a reference relating the system inputs and outputs. Therefore, the employment of a functional unit ensures that it is possible to present measures to improve one product (or system), or enables the comparison of two or more products [18, 19]. The functional unit must be clearly defined to describe quantitatively the function performed by any product, system process, or activity under analysis [23]. The selection of the functional unit should also be neutral to ensure that the specific quality and duration of the function is included. As an example, what is the functional unit that can be selected if an LCA study is used to compare different materials used in drinking cups (glass, plastic, ceramic)? The service provided by all the different cups needs to be specified, with enough information to allow the comparison, such as the quantity of the beverage for each cup, as an example 25 ml, and the lifetime of the cups (as the plastic cups can be used only once, and the ceramic or glass cups many times). Therefore, the functional unit can be defined as 1000 drinks (25 ml) or 1000 drinks (25 ml) per year. It is important to mention that some products or services can fulfil various functions (e.g. mobile phone, computer); and in these cases, one of the functions is chosen to represent the functional unit [16].

For the case study, the functional unit was defined as one small-calibre ammunition, as this functional unit enables the comparison between the four different ammunitions. This means that all data compiled, and the life-cycle results obtained, are related to that functional unit. However, selecting this type of functional unit narrows the possibility of relating the results from the LCA studies with other types of ammunition significantly different in concept (e.g. different calibre). If the goal of the LCA study is to compare the feasibility of the use of different ammunition types, it is necessary to provide a functional unit which delivers an appropriate indication of the ammunition function, such as the ammunition energy content (2,4,6-trinitrotoluene (TNT) equivalent) or its target efficiency. The same can be argued for the disposal of ammunition as the comparison of different demilitarization procedures (e.g. incineration, open detonation, sea dumping) need to be based on the energy content, or on the mass of ammunition demilitarized, instead of the number of ammunitions disposed.

7.2.1.2 System Boundaries

The principal objective of the system boundaries definition is intended to settle the borders of the assessment (what is included in the system under study). The definition of the boundaries follows an approach similar to that for environmental management, in which the system analysis takes into account the inputs and outputs associated with the processes/operations (e.g. manufacturing, transport, use/operation, and waste management) in a life-cycle perspective [18].

The system boundaries can comprise the entire life-cycle (cradle-to-grave), but can also be selected on the basis of the purpose of the study. For those studies, the boundaries are drawn in accordance with the objectives. Examples of system boundaries that only focus on a specific part of the life cycle of a product or system are cradle-to-gate – consider the life-cycle phases from the raw material extraction until the product leaves the gates of the factory; gate-to-grave (market studies) – the study begins after the manufacturing of the product; gate-to-gate (specialized unit process studies) – only studies the processes in the company; end-of-life/waste management.

The system boundaries should be defined in a way so that all processes are incorporated. However, that is not feasible in practice due to insufficient time, data, or resources. Usually, it is necessary to make some assumptions about which unit processes are included in the analysis, and the level of detail of that analysis [18]. These simplifications can be carried out with transparency and with appropriate justifications for the inclusion (or not) of these processes.

When selecting the system boundaries, and identifying its simplifications, it is useful to distinguish between 'foreground' and 'background' systems. The foreground system consists of those processes that are under the influence of the decision-maker and directly affects the study; whilst the background system comprises all the other processes that supplies energy, resources, and materials to the foreground system [16]. Differentiation between foreground and background systems is also important for deciding what type of data should be used. The foreground system should be described by specific process data, while the background is normally represented by average data from different technologies or processes [24]. This aspect also defines the direct impact nomenclature that is

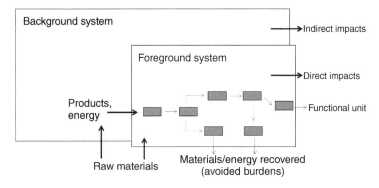

Figure 7.3 Representation of the foreground and background systems.

related to the foreground system, and the indirect impact nomenclature associated with the background system. The materials recycled or the energy recovered are credited as a negative input to the foreground system, and are usually named 'avoided burdens'. The total environmental burdens are calculated on the basis of the sum of direct and indirect burdens minus the avoided burdens if some type of impact is prevented. Figure 7.3 shows a representation of background and foreground systems as well as the connections between them that affect the calculation of the life-cycle impacts.

For the case study, the system boundaries include the processes for the production and use of the small-calibre ammunition. Figure 7.4 shows the life-cycle model of the production and use of small-calibre ammunition. The foreground system includes the production and use processes associated with the 9-mm small-calibre ammunition. The background system is made up of the processes that provide the materials and energy to the foreground system. These include the

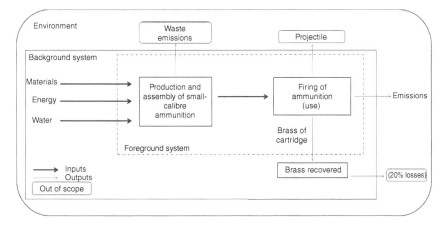

Figure 7.4 System boundaries associated with the production and use of small calibre ammunition. *Source:* Adapted from Ferreira et al. 2016 [22].

production of the ammunition materials (e.g. metals, energetic materials), water, and the production of energy based on the electricity mix for Portugal in the year 2004. The system boundaries also include the recovery of the ammunition cartridge after firing. The materials from the cartridge are incorporated in the production of, for example, new ammunition that displaces the use of raw materials, so the recycling part is accounted as avoided burdens.

Some assumptions referring to what is included (or not) in the analysis were performed because of method limitation or lack of data. Direct emissions associated with the production phase, transport between different life-cycle phases, and metal leaching from the projectile after firing are not included in the life-cycle model. With regard to the projectile which remains in the soil after firing, it was not possible to obtain information concerning the quantity and nature of metals which are eroded and leached. Those impacts are influenced by soil and weather characteristics and the LCA methodology does not address site-specific impacts, which can bias the assessment. Transport is also not included due to uncertain provenance of the materials and the locality of ammunition use. With regard to the cartridge, it is assumed that half is collected in the shooting ranges (50%) and the brass recovered displaces an equal quantity of virgin brass with losses of 20%.

Should the disposal of small-calibre ammunition be considered an important aspect for study, a new scenario that addressed that phase of the life cycle could be included. For instance, the system boundaries would include ammunition disposal, in which half the ammunition are used in training and the other half demilitarized.

7.2.2 Life-Cycle Inventory

The LCI analysis is the second phase of the LCA methodology, in which the data relevant to the system under study is collected. Collection and treatment of the data is one of the most time-consuming activities in LCA as it consists of performing an inventory of the numerical, descriptive, and qualitative data. The quantitative data consist of the collection of the inputs (e.g. raw materials, energy, components) and outputs (e.g. emissions, co-products) necessary to meet the goals of the defined study [16]. In addition, information regarding transport, equipment used in production, and other environmental aspects are also important. The data collected can be specific (e.g. from specific companies), but also more general (e.g. trade organizations, public surveys).

The process of data collection is iterative because new data can be needed as more is learnt about the processes, or new requirements or limitations are identified. In fact, the additional knowledge from the inventory can lead to the revision of the goal and scope of the study [16, 18]. Usually, the data used in the LCI is static in time (average data independent of time) and linear (the relationships are simplified to linear) [16]. The data collected also needs to be normalized in order to relate the inputs and outputs to the functional unit; therefore, data calculation is required such as conversion of units, relating activities to the main process, stoichiometric calculations, and allocation procedures [16].

The data collection (or measurement) is usually associated with the foreground processes dependent on the user or company that commissioned the study (named primary data). However, during the completion of the inventory it is not always possible to access all life-cycle data, principally data associated with processes that are not directly under the control of the practitioner (background processes), so that other sources need to be used. These data, named secondary data, can be available in scientific studies or technical reports published by organizations, companies, or consultants [16].

Another important source of information for background processes is the LCA software libraries that were created in order to simplify LCA studies and decrease the time spent in the compilation of data. The best known is the ecoinvent database that covers a broad range of life-cycle data and is published in extensive, transparent, and well-documented reports that can be accessed via the ecoinvent website (www.ecoinvent.org) or in LCA software. This data is relevant to different types of products (e.g. chemicals, metals, fuels, food) and activities (e.g. transport, energy production, waste treatment) and is updated regularly. These databases are the baseline for developing LCIs, but they also offer the possibility to change, adapt, or extend the databases to meet the needs of the intended study. In reality, if no data are available, the practitioner needs to fill the gaps with estimations, assumptions, or calculations (that need to be stated and justified in a transparent way) [16].

The inventory carried out to the production and use of the four small-calibre ammunition is detailed here. The LCI for the production and use phase of four types of 9-mm ammunition is based on primary data from the Romanian company U.M. Sadu – Gorj S.A. (production phase) and from Rotariu et al. [25] (use phase). The data is representative of similar small-calibre ammunition production in developed countries. Tables 7.1 and 7.2 show the materials for the production of the four different 9-mm ammunition.

The inventory includes the energy and water consumption associated with the assembly of the 9-mm ammunition. The information was obtained in a raw form (quantity of energy per year and the yearly production of small-calibre ammunition) that was normalized in accordance with the functional unit. It was also assumed that this consumption is equal for all the small-calibre ammunition as no data referent to the alternative materials was provided. Table 7.3 presents the energy and water requirement for the assembling of the ammunition, normalized to the functional unit (FU = 9 mm ammunition). It is important to mention that the data used for the electricity mix production and the production of some materials used in the production of parts of ammunition (such as metals, plastics, etc.) is based on the LCA databases (secondary data).

The inventories associated with the production of energetic materials are also included. As mentioned before, collection of data is the most time-consuming phase in an LCA study, and for military systems this difficulty is increased. The difficulty in obtaining data is significantly noticeable for the production of energetic materials, principally due to information confidentiality. Companies are not comfortable with providing data that can be covered by intellectual property or industrial secrecy, which leads to a scarcity of information regarding raw material, energy consumption, and emissions. In addition, the most common

Table 7.1 Data referent to the materials used for the production of 9-mm ammunition with a steel jacket and lead core projectile with lead primer (#1) and non-lead primer (#2).

Ammunition / Components	Steel/lead projectile + lead primer (#1) Constitution	Mass (kg)	Steel/lead projectile + non-lead primer (#2) Constitution	Mass (kg)
Cartridge	Brass	4.9E−03	Brass	4.9E−03
Projectile	Steel	3.9E−03	Steel	3.9E−03
	Lead	6.1E−03	Lead	6.1E−03
	Antimony	9.5E−05	Antimony	9.5E−05
Primer	Brass	2.4E−04	Brass	2.4E−04
	TNR-Pb	1.0E−05	DDNP	6.3E−06
	Tetrazen	1.3E−06	Tetrazen	1.3E−06
	Barium nitrate	4.9E−06	Zinc peroxide	1.4E−05
	Antimony sulfide	1.3E−06	Titanium powder	3.7E−06
	Lead dioxide	1.3E−06		
	Calcium silicide	1.3E−06		
Propellant	Single-base powder	4.1E−04	Single-base powder	4.1E−04
	Cardboard	3.2E−04	Cardboard	3.2E−04
Total weight		**1.6E−02**		**1.6E−02**

Source: Adapted from Ferreira et al. 2016 [22].

life-cycle databases (such as ecoinvent) do not cover this type of product, so data regarding the production and use of explosives is not available. In addition, details for the chemicals used in their production are limited, hindering the environmental assessment of explosives. In fact, the only information available in ecoinvent refers to a civil explosive for which the data on its production was obtained solely based on estimations as almost no real data were available [26] – this limitation was also tackled in Ferreira et al. [27] in which was assessed the environmental and toxicological impacts of the emulsion explosive production with the LCA methodology. Furthermore, the information in the literature is either scarce, rather outdated, or might be protected by military or trade secrecy.

The LCIs created by us for energetic materials were principally based on literature, covering information regarding the materials and energy used for production. Scarce literature information, or even for situations where the information was absent, meant that the LCIs were created on the basis of the procedure developed by Hischier et al. [28]. This approach suggests using an efficiency level of 95% for the stoichiometric chemical equation to account for raw materials consumption; the consumption of electricity and heat is based on average values (0.33 kWh and 2 MJ/kg of product) which are typical of the chemical industry [29]; and estimating that 0.2% of the input materials will be emitted

Table 7.2 Data referent to the materials used for the production of 9-mm ammunition with a composite projectile (nylon and copper) with lead primer (#3) and non-lead primer (#4).

Ammunition Components	Composite projectile + lead primer (#3)		Composite projectile + non-lead primer (#4)	
	Constitution	Mass (kg)	Constitution	Mass (kg)
Cartridge	Brass	4.9E−03	Brass	4.9E−03
Projectile	Nylon	4.1E−03	Nylon	4.1E−03
	Copper	1.0E−03	Copper	1.00E−03
Primer	Brass	2.4E−04	Brass	2.40E−04
	TNR-Pb	1.0E−05	DDNP	6.3E−06
	Tetrazen	1.3E−06	Tetrazen	1.3E−06
	Barium nitrate	4.9E−06	Zinc peroxide	1.4E−05
	Antimony sulfide	1.3E−06	Titanium powder	3.7E−06
	Lead dioxide	1.3E−06		
	Calcium silicide	1.3E−06		
Propellant	Single-base powder	4.1E−04	Single-base powder	4.1E−04
	Cardboard	3.2E−04	Cardboard	3.2E−04
Total weight		1.1E−02		1.1E−02

Table 7.3 Consumption of energy and water associated with the assembling of a small-calibre ammunition per functional unit – 9-mm ammunition.

Requirements for assembling	Amount (per 9-mm ammunition)
Electricity	0.046 kWh
Natural gas	0.067 kWh
Water	2.042 kg

Source: Adapted from Ferreira et al. 2016 [22].

into the air. However, these inventories can lead to an underestimation of the energy requirements, and also do not consider residues or waste emissions into water or soil, again due to lack of data. This can result in an incongruent mass balance (total mass of the used components might be higher than the total mass of the products).

For the case study, the ammunition producer provided primary data related to production of lead styphnate (TNR-Pb) and diazodinitrophenol (DDNP), whilst for tetrazene the data is taken from literature [30]. For the other five substances (barium nitrate, antimony sulfide, lead dioxide, zinc peroxide, and calcium silicide) the LCIs were created on the basis of the recommended approach mentioned above. As an example, Table 7.4 shows the inventory carried out for

Table 7.4 Life-cycle inventory for the production of RDX and the intermediary product hexamine.

Level 1	Level 2
RDX production (1 kg) Components[a]:	
Hexamine (0.8 kg)	Hexamine production (1 kg) Components[b]: Ammonia (0.51 kg); formaldehyde (1.3 kg) Emissions air[c]: Ammonia (0.001 kg); formaldehyde (0.002 kg) Energy: Electricity (0.35 kWh); steam (0.65 MJ)
Nitric acid (1.89 kg)	
Emissions air[c]:	
Nitric acid (0.0036 kg)	
Energy[c]:	
Electricity (0.333 kWh)	
Steam (2 MJ)	

a) Urbanski [30].
b) Aldehydes India Company.
c) Recommendations from Hischier et al. [28].

the production of 1, 3, 5-trinitrohexahydro-s-triazine (RDX) and the respective sources of information used. It is observed that to complete the inventory of RDX, it was also necessary to produce an inventory for hexamine production as no information regarding this chemical existed in the databases.

The collection of data associated with ammunition firing was performed following the study by Rotariu et al. [25], which provided information in more detail covering emission collection and analysis. Table 7.5 shows the emissions associated with the firing of the ammunition. These were also normalized to the functional unit (FU = 9 mm ammunition), in which the higher values are highlighted in bold for each emitted gas or metal. The analysis of the inventory regarding the firing emissions shows that the lead-free ammunition (principally ammunition #4), as expected, emits less lead. However, the ammunition with composite projectiles (#3 and #4) presents an increase in copper emissions. With regard to the gaseous emissions, ammunition #1 also produces higher emissions of NH_3, HCN, and CH_4. This data is insufficient to comprehend the potential effects of these emissions on the environment or on human health, so the next step of the LCA methodology (LCIA) needs to go further in the analysis in order to quantitatively assess the potential life-cycle environmental and toxicological consequences.

7.2.3 Life-Cycle Impact Assessment

The purpose of the LCIA phase is to provide additional information from the LCI, in which the inputs and outputs are converted into environmental consequences [19]. Therefore, the impact assessment characterizes and assesses

Table 7.5 Emissions associated with the firing of the four different 9-mm ammunition.

Substances	Emission (mg per ammunition)			
	Steel/lead projectile + lead primer (#1)	Steel/lead projectile + non-lead primer (#2)	Composite projectile + lead primer (#3)	Composite projectile + non-lead primer (#4)
CO	198.65	184.75	119.21	118.76
CO_2	101.79	96.79	58.56	57.93
NO	3.80	3.22	3.85	**4.41**
NO_2	0.64	0.62	0.49	0.52
NH_3	3.10	2.46	1.67	1.84
HCN	1.77	1.22	0.18	0.13
CH_4	1.10	0.96	0.61	0.59
Pb	3.14	1.04	0.81	0.04
Cu	0.55	0.41	4.85	**5.21**
Zn	0.12	0.11	**0.19**	0.03
Sb	0.37	0.20	0.15	Not detected

Source: Adapted from Ferreira et al. 2016 [22].

quantitatively the effects on the environment and the human health associated with the mass and energy fluxes identified in the inventory. The LCIA phase comprises two mandatory steps (classification and characterization) and two optional steps (normalization and weighting). It is important to mention that the LCIA methods have already carried out the classification and characterization steps, so the users are only required to select the methods, and the environmental impact categories, that are intended to be used.

The classification step involves the sorting and assignment of the inventory data to specific environmental impact categories and category indicators [16, 19]. Classification is a qualitative step based on scientific analysis of relevant environmental mechanisms [18]. Certain outputs (e.g. emissions) contribute to various impact categories, which are assigned to different impact categories (e.g. methane contributes to global warming effects and photo-chemical oxidation effects).

Characterization is a quantitative step based on the natural sciences, relating the physicochemical mechanism of each substance that influences its behaviour for each impact category [16]. The characterization has to assign the relative contribution of each input and output to the selected impact categories with the assistance of specific factors. For some of the environmental impact categories, there is consensus about the factors to be used in the estimation of the total impact (e.g. global warming, ozone depletion); whilst for other environmental impacts, the consensus for the conversion factors are still controversial (e.g. biotic resources, toxicity, land use) [16, 18]. The impact characterization can be divided into two approaches that link the inventory data and the potential environmental impacts:

- *Problem oriented:* Stop the quantitative modelling before the end of the impact pathway that links the inventory to the inherent environmental problems (midpoint categories) such as acidification or ozone depletion. The midpoint results have the advantage of a lower uncertainty associated with the cause–effect chain between the inventory and the potential impacts; however, the results are more difficult to understand by the general population, principally an audience that is not from the environmental field.
- *Damage oriented:* The calculation of the cause–effect chain goes up to the environmental damages (endpoint categories). These results increase the effectiveness of transmitting the results and the message with a broader audience, but increase the uncertainty as it is difficult to predict the pathway that leads to the final consequence from the initial interventions.

The connections between the inventory, the environmental impact (problems), and its potential damages are presented in Figure 7.5. It can be argued that environmental impacts will always lead to a consequence (damage). However, the evaluation of the damage impact is not mandatory in an LCA study due to the uncertainty associated with the evaluation of the consequences.

The calculation of environmental impact is associated with the environmental mechanism mentioned earlier that calculates the magnitude of the potential consequences for each substance, named as characterization factors (CFs). The CFs are developed on the basis of the properties of substances for the specific impact

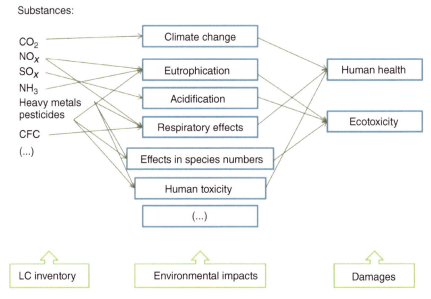

Figure 7.5 Representation of the connection between the inventory data, environmental impact categories (problem oriented), and effects in the human health and ecosystems (damage oriented).

categories. The CFs for each impact category are related to a selected substance that is used as a reference. For instance, for Climate Change the CFs are modelled on the basis of the Intergovernmental Panel on Climate Change (IPCC), in which CO_2 is the reference substance. Therefore, the CF for CO_2 in the impact category Climate Change has the value of 1, while the CF of methane is 25. This means that the release of 1 kg methane causes the same amount of impact as an emission of 25 kg of CO_2.

The other two optional steps in LCIA are briefly described. Normalization can be used to provide information on the importance (magnitude) of the impact of a product under study [19]. Normalization relates the LCIA results associated with a certain product or service by dividing it with the impacts associated with a reference activity or situation [31]. The activity whose impacts have been selected by policy makers as a reference – named normalization factors – is usually the regional economic system [32]. That means the normalization factors for each impact category are the yearly impacts associated with the emissions and extraction data of a certain regional or global economic activity. Weighting is a procedure that can be quantitative or qualitative and applies weights (weighting factors) to each impact category in accordance with the relevance attributed by, for instance, decision-makers or specialists [16].

7.2.3.1 Life-Cycle Impact Assessment Methods

The most general LCIA methods are the CML (developed by the institute of the Faculty of Science – CML – of Leiden University) and ReCiPe, which proposed a set of impact categories and CFs for the impact assessment phase [33]. These two LCIA methods, going beyond the environmental impact categories, have toxicological categories that are not usually included in the analysis as they involve a high degree of uncertainty and lack scientific robustness [34]. In the case study, the CML method was employed to calculate the life-cycle environmental impacts, so Table 7.6 presents the description of six environmental impact categories selected from that method.

Assessment of toxicity impacts on human health and ecosystems are of central importance for LCA studies. However, different LCIA methods usually fail to arrive at the same toxicological impact for a certain substance [36, 37]. Therefore, toxicity issues are not typically addressed in LCIA. The high uncertainty related with the toxicological CFs (approximately 12 orders of magnitude) also contributed to neglect of the toxicity issues [38].

Recently, the USEtox method was developed with the aim of improving the toxicological assessment in LCA studies. In 2008, the USEtox method was recommended by UNEP/SETAC (United Nations Environment Programme/Society of Environmental Toxicology and Chemistry), in a first-time consensus on how to calculate toxicological CFs in LCA. The consensus was obtained essentially due to (i) the largest substance coverage presently available; (ii) the transparent calculation of CFs; (iii) and the low estimated uncertainty (100–1000 for human health, and 10–100 for freshwater ecotoxicity) in comparison with previous LCIA methods [38]. The USEtox impact categories were also recommended by the European Commission's Joint Research Centre to assess toxicological impacts [39].

Table 7.6 Description of six environmental impact categories from CML method.

Impact category	Description	Unit
Abiotic depletion (AD)	Extraction of minerals and fossil fuels, based on concentration of reserves and rate of de-accumulation	kg Sb equiv
Acidification (Acid)	Describes the fate and deposition of acidifying substances	kg SO_2 equiv
Eutrophication (Eut)	Impacts due to excessive levels of macronutrients in the aquatic environment	kg PO_4^{3-} equiv
Global warming (GW)	Emissions of greenhouse gases	kg CO_2 equiv
Ozone layer depletion (OLD)	Emission of gases that have an effect in the destruction of the stratospheric ozone layer	kg CFC-11 equiv
Photochemical oxidation (OP)	Creation of reactive substances (mainly ozone)	kg C_2H_4 equiv

Source: Adapted from Goedkoop et al. 2010 [35].

The USEtox method calculates the CFs for human toxicity (cancer and non-cancer) and ecotoxicity based on fate, exposure, and effect factors [38]. The three factors aforementioned are calculated on the basis of the physic-chemical and toxicity properties as well as the environmental characteristics that influence the behaviour of substances on the environment. The fate factor calculates the dispersion and transfer of a chemical between the different compartments (air, water, and soil). The exposure factor determines the transfer of a chemical from an environmental media into humans and/or ecosystem system accordingly to different exposure pathways (inhalation and ingestion – drinking water, meat, fish, dairy products, etc.). The effect factor accounts for the toxicity on humans and different trophic organisms (e.g. plants, microorganisms, algae, fish) [38]. Therefore, the USEtox method converts the physico-chemical and toxicity properties and environmental characteristics into potential impacts for ecosystems and human health. Figure 7.6 shows the framework of USEtox to calculate the toxicological characterization factors that are employed to assess the potential toxicological impacts.

USEtox provides the possibility of selecting where the emission will occur. Among the different options, USEtox considers seven environmental compartments: indoor air, urban air, continental rural air, freshwater, sea water, natural soil, and agricultural soil. The selection of an emission compartment is necessary to determine the partition of a substance between different media (e.g. air, soil). Each one of these media has its own environmental characteristics that will determine the behaviour of a substance in the environment (the fate and exposure). The USEtox method also allows choosing between a global or continental scale.

The toxicological categories are expressed in comparative toxicity units for human health (CTUh) and ecosystems (CTUe). Rosenbaum et al. [38] described

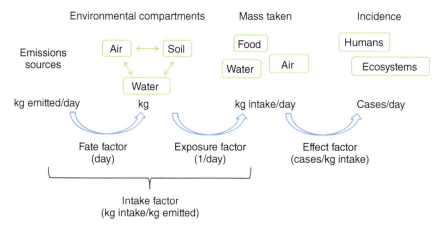

Figure 7.6 USEtox framework for calculating the toxicological impacts on human health and ecosystems. *Source:* Adapted from Fantke et al. 2015 [40].

in detail the units for CFs, defining human toxicity as 'provides the estimated increase in morbidity cases in the total human population per unit mass of a chemical emitted' and ecotoxicity 'provides an estimate of the potentially affected fraction (PAF) of species integrated over time and volume per unit mass of a chemical emitted'.

Despite the consensus that USEtox reached, the method still contains some concerns associated with uncertainty. The USEtox method provides two sets of CFs (recommended and indicative) based on the level of reliability of the toxicological assessment [40]. The recommended CFs are the ones that have reached a scientific consensus, whilst the indicative CFs are classified as provisional due to the relatively higher uncertainty [38, 40]. Metals and amphiphilic and dissociating substances are classified as indicative in USEtox, so the use of CFs of these substances should be done with great caution due to the inherent uncertainty. The uncertainty is due to the inability of the model to consider the complex behaviour of substances in the environment or due to the lack of available substance data [40].

7.2.3.2 Life-Cycle Impact Assessment Software

The environmental impacts are calculated with the employment of software that helps and facilitates the modelling of the processes under study. One package is SimaPro which can be applied to the modelling of both products and systems from a life-cycle perspective in order to calculate the potential environmental impacts. With this software, it is possible to build complex models in a systematic and transparent way using the databases which are fully integrated in the software. SimaPro contains LCI databases with a broad international scope, including the comprehensive Swiss-based ecoinvent database. All databases are completely harmonized regarding structure and nomenclature and fit well in SimaPro with all impact assessment methods. A variety of LCIA methods are also embedded in the software, where it is possible to select which one to use. SimaPro has a structure based on the four phases of LCA, with all the information describing the process and products in the databases available in terms of input and output flows of substances. Processes can be linked to each other to create networks, and it is also possible to copy and update the information (inputs and outputs) to specific processes.

7.2.3.3 Life-Cycle Impact Assessment of the Case Study

As mentioned, the Simapro package was used to model and calculate the life-cycle environmental and toxicological impact with the CML and USEtox methods, respectively. Firstly, the most relevant life-cycle phases (production or use phase) of the small-calibre ammunition for environmental and toxicological impacts were determined. Secondly, the environmental and toxicological impacts were presented, and the contributors to those impacts were identified and discussed. Finally, some procedures to improve the environmental performance based on the results obtained for those ammunition were presented.

Which is the most important life-cycle phase for the small-calibre ammunition case?

Figure 7.7 presents the environmental impacts calculated with the CML method for the three life-cycle phases considered in this study: production, use, and brass recovery (cartridge). In order to facilitate the representation and comparison of the impact for each category (which uses different units), the results obtained are transformed in values between 0 and 1 (1 being the highest impact value for each impact category). The production phase gives a higher contribution to the environmental impact categories, representing more than 90% of the impacts. The use phase has only a significant contribution to the photo-chemical oxidation (≈10%). The brass recovered from the cartridge presents a positive

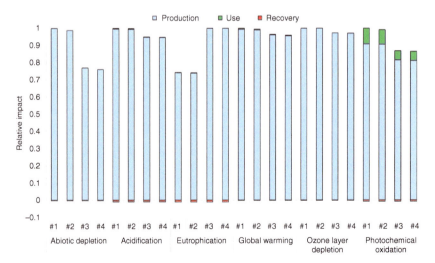

Figure 7.7 Environmental life-cycle impacts, calculated with the CML method, associated with the production, use and recovery of brass from four different small calibre ammunition: #1: steel/lead projectile + lead primer; #2: steel/lead projectile + non-lead primer; #3: composite projectile + lead primer; #4: composite projectile + non-lead primer. *Source:* Adapted from Ferreira et al. 2016 [22].

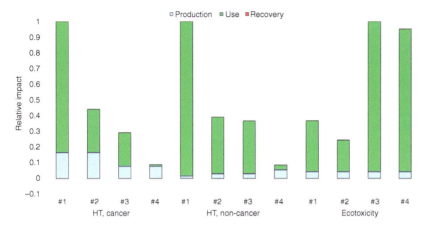

Figure 7.8 Toxicological life-cycle impacts, calculated with the USEtox method, associated with the production, use and recovery of brass from four different small calibre ammunition: #1: steel/lead projectile + lead primer; #2: steel/lead projectile + non-lead primer; #3: composite projectile + lead primer; #4: composite projectile + non-lead primer.
Source: Adapted from Ferreira et al. 2016 [22].

impact for all environmental impact categories (which are represented as negative values due to the avoidance of environmental impacts associated with the displacement of virgin brass); however, the benefit from the recovery of brass has a low significance to the overall life-cycle impact.

Figure 7.8 presents the toxicological life-cycle impacts calculated with the USEtox method to the three life-cycle phases considered in this study (production, use, and brass recovered from the cartridge), in which the analysis carried out is similar to that for the environmental impact categories. The results show that the use phase dominates the total life-cycle impacts, contributing to more than 80% to the majority of the small-calibre ammunition. The only exception is for the Human Toxicity categories as for ammunition #4 the production phase presents a higher contribution to the total impact.

A detailed analysis of the results enables us to answer some questions that are addressed in the next paragraphs.

- Which are the small-calibre ammunition with lower environmental and toxicological impact?

It is not clear which ammunition produces less environmental impact based on the six environmental categories. For three out of six categories (Acidification, Global Warming, and Ozone Layer Depletion) the results for ammunition #3 and #4 shows only slightly lower impact (approximately less than 5%) than the ones obtained for ammunition #1 and #2. Nevertheless, ammunition #1 and #2 present significantly higher impacts for Abiotic Depletion (23%) and Photo-chemical Oxidation (13%); while the ammunition with a composite projectile (#3 and #4) produces a higher impact for Eutrophication (26%).

The analysis of the toxicological impacts helps comprehend which ammunition produces lower impacts. The substitution of lead offers a potential toxicological benefit on human health as the small-calibre ammunition without lead has a lower impact on human health. In fact, substitution of lead in the primer and projectile (ammunition #4) almost completely avoided the impacts associated with the use phase. Just the removal of lead from the primer allowed a reduction of 56% (cancer effects) and 61% (non-cancer effects) of the total toxicological life-cycle impact (ammunition #1 vs ammunition #2), while the use of the composite projectile in comparison with the steel/lead projectile (ammunition #1 vs ammunition #3) allowed an impact reduction of 71% (cancer effects) and 63% (non-cancer effects). However, the analysis of the ecotoxicity impact category shows a trade-off as the ammunition with a composite projectile presents around 65% higher life-cycle impacts than the small-calibre ammunition with a steel/lead projectile.

Due to the trade-off observed for the toxicological impact categories (as well as for the environmental impact categories), the selection of the appropriate small-calibre ammunition is difficult. Therefore, the emphasis given by decision-makers to the various impact categories (for instance, human health being more important than environmental impact categories) is what can influence the decision on which ammunition should be selected.

After the analysis of these results, some questions still need to be answered. It is important to understand why the production phase has a higher contribution to the environmental impact categories, while the use phase has a higher contribution to the toxicological impact categories. Moreover, it is necessary to understand the differences observed for the four different small-calibre ammunition types. In order to answer those questions, it is necessary to identify what is contributing to the life-cycle impacts, and offer recommendations to decrease those observed impacts. With the LCIA, is possible to go into this type of detail and comprehend what is causing those impacts.

- What are the main contributors to the life-cycle impacts?

Figure 7.9 shows the main contributors to the environmental and toxicological life-cycle impact associated with production phase for ammunition #1 and #3 (as ammunition with the same type of projectile present similar impact contribution). The results are shown as a percentage of the total production phase impact for each one of the categories, with the aim of determining what is the contribution of the main activities: components (projectile, cartridge, primer, and propellant) and energy consumption.

The cartridge, the propellant, and the energy requirement are equal factors for all ammunition, so the overall contribution to the production impact of these components and energy is constant. Consequently, the differences observed between the four small-calibre ammunition arise from the alternative projectiles and primers. However, as the relative contribution to the impact associated with the projectile and primer can vary, the relative impact contribution of the components can also change.

The main contributors to the six environmental impact categories are the energy consumption and the cumulative embodied impacts associated with the materials used in the production of the projectile and the cartridge. As the cartridge and the energy requirement are the same for all ammunition (as mentioned earlier), and the impact associated with the production of the

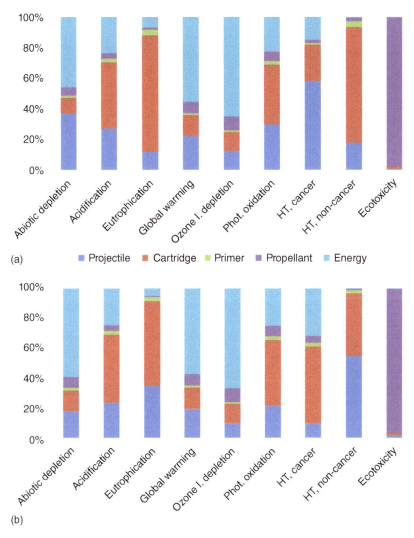

Figure 7.9 Main contributors to the environmental life-cycle impacts, calculated with the CML method, associated with the production of the small calibre ammunition #1 and #3. (a) Steel/lead projectile + lead primer (#1); (b) Composite projectile + lead primer (#3). *Source:* Adapted from Ferreira et al. 2016 [22].

primer is minor to the overall production impact, the variation observed between the four ammunition types can only be associated with the differences in the projectile.

The contribution of the steel/lead projectile (ammunition #1) to the impact categories Abiotic Depletion ($\approx$38%) and Photo-chemical Oxidation ($\approx$35%) is higher compared to the contribution of the composite projectile (ammunition #3) to those impact categories – Abiotic Depletion ($\approx$18%) and Photo-chemical Oxidation ($\approx$21%). The indirect impacts associated with the production of steel and lead has a higher impact for these two impact categories than for the indirect impacts from the components of the composite projectile (copper and nylon), resulting in the differences observed. That is also the reason for the higher impacts of the composite projectile for Eutrophication, in which the production of nylon has a higher contribution to this impact category than to the production of lead and steel. The determination of these results, and the transparent identification of the main influential activities (or products) on the life-cycle impact, is possible due to the employment of a life-cycle approach as the background burdens (indirect impacts) associated with energy and material production are also considered.

Another important aspect which it was possible to detect with the employment of a life-cycle approach for the production phase was the identification of the main contributor to the ecotoxicity impact category. Propellant production dominated the environmental life-cycle results with a contribution higher than 90% for all the ammunition types. A detailed analysis of this component allows the determination of the toxicological impact for ecosystems principally associated with the emissions of pesticides used in the cultivation of cotton (which is one of the main components for the production of nitrocellulose). This conclusion is only possible with the inclusion of a life-cycle perspective that enables the inclusion of the background activities, which can be important for the assessment in order to achieve a sustainable approach.

Figure 7.10 shows the contributors to the toxicological impact associated with ammunition firing during the use phase. The analysis is carried out in the same way as the one for the production phase, in which the results are presented as a percentage of the total use phase impact for each one of the toxicological categories. Lead dominates the impact contribution for the Human Toxicity categories, and the reduction in quantity of this substance has a direct consequence in the impact decrease. This is the reason for the lower toxicological impact on human health for the small-calibre ammunition without lead in the primer or projectile (or both), as shown in Figure 7.8. In fact, for the ammunition with composite projectiles, the lead emissions are decreased to a point that zinc also presents a significant contribution to the use phase impact.

For the ecotoxicity impact category, the high impact observed for the ammunition with composite projectiles is the consequence of an impressive increase of copper emissions; whilst the main contributor to the ecotoxicity impact for the ammunition with steel/lead projectile is antimony (approximately 60%). Copper has a higher effect on ecosystems in comparison with antimony, resulting in the differences between the four ammunition under study. In fact, the effect of copper on human health is also of great importance as this substance can cause metal fume fever.[1] Moxnes et al. [41] stated that military personal that used similar ammunition with a composite projectile remained out of action for several hours due to the acute effect of this substance.

Based on the results, some overall recommendations can be proposed to improve the environmental performance of small-calibre ammunition. The main

---

1 The intense exposure to different varieties of metals can originate symptoms that are similar to the ones resulting from influenza, such as fever, sweating, myalgia, etc.

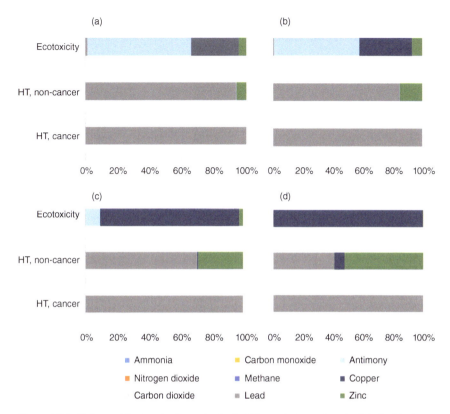

Figure 7.10 Main contributors to the toxicological life-cycle impacts, calculated with the USEtox method, associated with the production of the four small calibre ammunition. (a) Steel/lead projectile + lead primer (#1); (b) Steel/lead projectile + non-lead primer (#2); (c) Composite projectile + lead primer (#3); and (d) Composite projectile + non-lead primer (#4). *Source:* Adapted from Ferreira et al. 2016 [22].

impact for production is associated with the energy consumption and of the cartridge, so one important measure proposed is to reduce the energy required, and substitute the brass cartridge for cardboard or eventually, the use of caseless ammunition. As some burdens also result from the projectiles, and the substitution of lead in the projectile seams an appropriate measure to decrease impact on human toxicity, it is important to propose composite projectiles with alternative materials instead of copper-nylon, such as steel-nylon (or other potential combinations) and thus avoid the impact on Eutrophication (or other potential trade-offs).

It is important to emphasize that the conclusions drawn regarding the substitution of lead from the ammunition components are due to the high chronic effects on human health. A decrease of the potential impact on human health is observed when free lead materials are used in the primer and in the projectile. This might be an appropriate solution to apply in shooting ranges during training. Nevertheless, this approach should be carried out with caution in order to prevent the shifting of the toxicological impacts origin to other metals (such as copper, zinc, and antimony). Therefore, a life-cycle approach (and ecodesign studies), such as the one presented, is required to support such decisions.

7.3 Limitations of Life-Cycle Assessment

LCA allows for the calculation of environmental impacts, but there are some limitations that should be stressed. The principal characteristic of LCA is its holistic nature, which is its main strength and, at the same time, its limitation. Achieving the broad scope of examining the complete life cycle of a product is only made possible by simplifying or making assumptions for many aspects. LCA is typically a steady-state rather than a dynamic approach, and generally considers that all processes are linear. This methodology focuses on environmental aspects and usually does not consider economic, social, and site-specific impacts. Moreover, LCA is an analytical tool that provides information for decision support, but it cannot replace the decision making process itself. That is to say, LCA is a tool to assist the decision but it is not the decision itself. In fact, LCA is a complementary tool that provides additional knowledge to other types of environmental management methodologies, such as risk analysis or environmental impact assessment. The data available can also frequently be obsolete, incomparable, or of unknown quality.

Beyond these limitations regarding the LCI, LCIA methods also lack CFs for some substances and their environmental consequences. This limitation is even greater for the toxicological impact categories. The lack of toxicological CFs is usually associated with the high range of substances that are not included in the methods; lack of available real data regarding the substance properties (e.g. toxicity dose–response) which need to be evaluated by costly experiment; and difficulties in modelling and predicting the complex behaviour of substances in the environment. However, this limitation is less critical since the creation of the USEtox method, through which it is possible to calculate the toxicological CFs for substance not yet available in the databases.

7.4 Conclusions

This chapter discusses in detail LCA methodology, and attempted to show the capability and the type of information that it is possible to obtain when this methodology is employed for military systems. The case study that was used to illustrate the four phases of the LCA methodology is relevant to the comparison of the production and use of four different types of 9-mm ammunition. This study demonstrated that LCA can be used to (i) understand which life-cycle phases produce higher impact; (ii) determine the main contributors to the life-cycle environmental and toxicological impacts of a product; (iii) compare products with the same function; and (iv) produce recommendations to decrease the environmental and toxicological burdens.

Conclusions drawn from LCA studies are relevant for stakeholders, decision-makers, range managers, and manufacturers concerned with environmental issues related to ammunition. Based on this type of assessment, it is possible to begin to quantify and assess the main environmental and toxicological concerns, and provide tailored and robust decisions in order to mitigate (or decrease) the impacts in a life cycle.

References

1 Clausen, J. and Korte, N. (2009). The distribution of metals in soils and pore water at three U.S. military training facilities. *Soil Sediment Contam.* 18: 546–563.
2 Hewitt, A.D., Jenkins, T.F., Walsh, M.E. et al. (2005). RDX and TNT residues from live-fire and blow-in-place detonations. *Chemosphere* 61: 888–894.
3 Perroy, R.L., Belby, C.S., and Mertens, C.J. (2014). Mapping and modelling three dimensional lead contamination in the wetland sediments of a former trap-shooting range. *Sci. Total Environ.* 487: 72–81.
4 RTO (Research and Technology Organization) (2010) Environmental Impact of Munition and Propellant Disposal. *Final Rep. Task Group AVT-115*. North Atlantic Treaty Organization.
5 RTO (Research and Technology Organization) (2014) Design for Disposal of Present and Future Munitions and Application of Greener Munition Technology. *Final Rep. Task Group AVT-179*. North Atlantic Treaty Organization.
6 RTO (Research and Technology Organization) (2015). Munitions-Related Contamination – Source Characterization, Fate and Transport. *Final Rep. Task Group AVT-197*. North Atlantic Treaty Organization.
7 Thiboutot, S., Brousseau, P., and Ampleman, G. (2015). Deposition of PETN following the detonation of seismoplast plastic explosive. *Propellants Explos. Pyrotech.* 40: 329–332.
8 Walsh, M.R., Walsh, M.E., and Ramsey, C.A. (2012). Measuring energetic contaminant deposition rates on snow. *Water Air Soil Pollut.* 223: 3689–3699.
9 Walsh, M.R., Walsh, M.E., Ramsey, C.A. et al. (2013). Perchlorate contamination from the detonation of insensitive high-explosive rounds. *J. Hazard. Mater.* 262: 228–233.
10 Walsh, M.R., Walsh, M.E., Ampleman, G. et al. (2014). Munitions propellants residue deposition rates on military training ranges. *Propellants Explos. Pyrotech.* 37: 393–406.
11 Fisher, I.J., Pain, D.J., and Thomas, V.G. (2006). A review of lead poisoning from ammunition sources in terrestrial birds. *Biol. Conserv.* 131: 421–432.
12 Helander, B., Axelsson, J., Borg, H. et al. (2009). Ingestion of lead from ammunition and lead concentrations in white-tailed sea eagles (*Haliaeetus albicilla*) in Sweden. *Sci. Total Environ.* 407: 5555–5563.
13 Bonanno, J., Robson, M.G., Buckley, B., and Modica, M. (2002). Lead exposure at a covered outdoor firing range. *Bull. Environ. Contam. Toxicol.* 68: 315–323.
14 Pichtel, J. (2012). Distribution and fate of military explosives and propellants in soil: a review. *Appl. Environ. Soil Sci.* 2012: 1–33.

15 Ferreira, C., Ribeiro, J., Mendes, R., and Freire, F. (2013). Life-cycle assessment of ammunition demilitarization in a static kiln. *Propellants Explos. Pyrotech.* 38: 296–302.
16 Baumann, H. and Tillman, A.M. (2004). An orientation in life cycle assessment methodology and application. In: *The Hitch Hiker's Guide to LCA*. Lund, Sweden: Studentlitteratur. ISBN: 9144023642.
17 European commission – Joint Research Center (2010). A guide for business and policy makers to life cycle thinking and assessment. In: *Making Sustainable Consumption and Production a Reality*. Luxemburg: Publications Office of the European Union. ISBN: 978-92-79-14357-1.
18 Jensen, A., Hoffman, L., Møller, B.T. et al. (1997). A guide to approaches, experiences and information sources. In: *Life-Cycle Assessment (LCA)*, Environmental Issues Series, vol. 6. Denmark: dk-TEKNIK Energy and Environment.
19 ISO 14040 (2006) Environmental management – life cycle assessment – principles and framework. International Organization for Standardization, Geneva.
20 ISO 14044 (2006) Environmental management – life cycle assessment – requirements and guidelines, Geneva.
21 Monteiro, H. and Freire, F. (2012). Life-cycle assessment of a house with alternative exterior walls: comparison of three impact assessment methods. *Energy Build.* 47: 572–583.
22 Ferreira, C., Ribeiro, J., Almada, S. et al. (2016). Reducing impacts from ammunitions: a comparative life-cycle assessment of four types of 9 mm ammunitions. *Sci. Total Environ.* 34–40: 566–567.
23 Rebitzer, G., Ekvall, T., Frischknecht, R. et al. (2004). Life cycle assessment part 1: framework, goal and scope definition, inventory analysis, and applications. *Environ. Int.* 30: 701–720.
24 Azapagic, A. and Clift, R. (1999). Allocation of environmental burdens in multiple-function systems. *J. Cleaner Prod.* 7: 101–119.
25 Rotariu, T., Petre, R., Zecheru, T. et al. (2015). Comparative study of 9 × 19 mm ammunition combustion products and residues. *Propellants Explos. Pyrotech.* 40: 931–937.
26 Kellenberger, D., Althaus, H., Künniger, T., Lehmann, M. (2007). Life Cycle Inventories of Building Products. *Ecoinvent Rep. No. 7*, Dübendorf.
27 Ferreira, C., Freire, F., and Ribeiro, J. (2015). Life-cycle assessment of a civil explosive. *J. Cleaner Prod.* 89: 159–164.
28 Hischier, R., Hellweg, S., Capello, C., and Primas, A. (2005). Establishing life cycle inventories of chemicals based on differing data availability. *Int. J. Life Cycle Assess.* 10: 59–67.
29 Gendorf (2000). *Umwelterklärung*. Burgkirchen: Industrial Park Werk Gendorf.
30 Urbanski, T. (1968). *Chemistry and Technology of Explosives*, vol. 1–4. Oxford: Pergamon Press.
31 Kim, J., Yang, Y., Bae, J., and Suh, S. (2013). The importance of normalization references in interpreting life cycle assessment results. *J. Ind. Ecol.* 17 (3): 385–395. https://doi.org/10.1111/j.1530-9290.2012.00535.x.

32 Sleeswijk, A.W., van Oers, L.F.C.M., Guinée, J.B. et al. (2008). Normalisation in product life cycle assessment: an LCA of the global and European economic systems in the year 2000. *Sci. Total Environ.* 390: 227–240.

33 Guinée, J.B., Gorrée, M., Heijungs, R. et al. (2001). Operational guide to the ISO standards. I: LCA in perspective. IIa: Guide. IIb: Operational annex. III: Scientific background. In: *Handbook on Life Cycle Assessment*. Dordrecht: Kluwer Academic Publishers.

34 Finnvenden, G., Hauschid, M.Z., Ekvall, T. et al. (2009). Recent developments in life-cycle assessment. *J. Environ. Manage.* 91: 1–21.

35 Goedkoop, M., Oele, M., Schryver, A. de et al. (2010) SimaPro Database Manual Methods Library. *Rep. Version 2.4*, PRé Consultants, The Netherlands.

36 Caneghem, J.V., Block, C., and Vandecasteele, C. (2010). Assessment of the impact on human health of industrial emissions to air: does the result depend on the applied method? *J. Hazard. Mater.* 184: 788–797.

37 Mattila, T., Verta, M., and Seppälä, J. (2011). Comparing priority setting in integrated hazardous substance assessment and in life cycle impact assessment. *Int. J. Life Cycle Assess.* 16: 788–794.

38 Rosenbaum, K., Bachmann, M., Gold, S. et al. (2008). USEtox – the UNEP-SETAC toxicity model: recommended characterisation factors for human toxicity and freshwater ecotoxicity in life cycle impact assessment. *Int. J. Life Cycle Assess.* 13 (7): 532–546.

39 European commission – Joint Research Center (2011). Recommendation for life cycle impact assessment in the European context – based on existing environmental impact assessment models and factors. In: *ILCD Handbook*, 1ee. Luxemburg: Publications Office of the European Union.

40 Fantke, P.E., Bengoa, X., Dong, Y. et al. (2015) USEtox® 2.0 Documentation, USEtox® Team, Kgs. Lyngby.

41 Moxnes, J.F., Jensen, T.L., Smestad, E. et al. (2013). Lead free ammunition without toxic propellant gases. *Propellants Explos. Pyrotech.* 28: 255–260.

8

Integrating the 'One Health' Approach in the Design of Sustainable Munition Systems

Mark S. Johnson

U.S. Army Public Health Center, Toxicology Directorate, 8258 Blackhawk Road, Aberdeen Proving Ground, Maryland

8.1 General Background

8.2 Munition Compounds and Aetiology of Environmental, Safety, and Occupational Health Issues: Lessons Learnt

The concept of life-cycle analysis involves the cradle-to-grave consideration of all costs associated with the development of a new compound or system. Increasingly, unexpected costs associated with production and use have halted training activities, resulted in increased production costs and delays, and affected the health of warfighters and workers exposed to these materials. In many cases, substitutions are not available and means to reduce exposures are the only options. Hazardous wastes are accumulated in production facilities and must be disposed of in costly permitted landfills. Many substances were developed and used with only a minimal investment in the necessary toxicology data needed to allow decision-makers to make informed and balanced decisions. Hence, accurate life-cycle costs were not captured or debated.

The use of the oxidizer, ammonium perchlorate, was judged to be acceptable based on its relatively low acute toxicity. However, low-level, long-term exposures have been shown to affect thyroid hormone production through inhibition of the sodium-iodide symporter affecting iodide uptake [1]. This mechanism was discovered long after the production and use of perchlorate. Moreover, the perchlorate anion is very water-soluble and was found to infiltrate soils at production and testing facilities and subsequently found in some groundwater resources. This has become an issue of national importance in the United States, and research into alternatives has begun.

The explosive, 1,3,5-hexahydro-1,3,5-trinitrate (RDX) has also been found in groundwater resources and has been found to migrate off military installations affecting drinking water sources [2]. In some cases, training operations have been halted by states where contaminated groundwater has migrated off instal-

Energetic Materials and Munitions: Life Cycle Management, Environmental Impact and Demilitarization,
First Edition. Edited by Adam S. Cumming and Mark S. Johnson.
© 2019 Wiley-VCH Verlag GmbH & Co. KGaA. Published 2019 by Wiley-VCH Verlag GmbH & Co. KGaA.

lations. Pink water has been found historically at 2,4,6-trinitrotoluene (TNT) manufacturing sites, resulting in discharges to soil [3]. The use of coloured smoke canisters have been found to cause toxicity to soldiers when used as an obscurant in urban fighting situations, not when used as intended as a signalling tool.

Development of toxicology data must not be a costly endeavour and should be conducted in a phased manner coincident with the level of investment devoted to new compound development and weapons systems design. The use of the phased approach to gathering these data where environmental, safety, and occupational health (ESOH) testing is accomplished alongside research, development, testing, and evaluation can provide results that can maintain mission requirements and keep projects and costs on schedule.

The "One Health" concept considers all potential linkages of exposure and effects to include those that occur from environmental releases to non-human entities and how that may subsequently affect public health and other environmental processes. It includes fate and transport of parent compounds, combustion, and environmental breakdown products and how exposure may affect ecosystem processes and human health. This vision is consistent with green chemistry approaches and full life-cycle assessment.

8.3 Core Operational ESOH Data: Needs and Requirements

8.3.1 Life Cycle Environmental Assessment

Specific data are needed to ascertain degree of risk and for specific tools to be used to ensure safe and sustained use. The US Army has regulations (AR 40-5, AR 70-1) that require program managers for weapons systems and platforms consider and integrate ESOH concerns into full life-cycle considerations; other allied nations have similar procedures and considerations. However, rarely are specific toxicology data requirements specified. Documents to include the Programmatic Environmental Safety and Health Evaluation (PESHE) and the Life Cycle Environmental Assessment are requirements, although they are often lacking descriptive data necessary to ensure safe operation and enable sustained use.

8.3.2 Bridging Communication Between Research and Acquisition

Military research for new weapons systems can involve everything from new compound development, to understanding the impact of new formulations and additives, and engineering of systems. If successful, through pure research, applied research, demonstration of proof-of-principle, and testing, the work is optimally transitioned to acquisition programs that will implement new and emerging technologies into the production of a new system. The responsible entity of this new technology becomes the program manager who must understand the full life-cycle costs associated with implementation. Acquisition programs require many levels of oversight and areas of consideration. Program managers are best served by accepting new technologies with sufficient ESOH information and when recommendations are provided. In US Army programs, a Toxicity Assessment is a technical foundation that contains toxicity information on substances used in new systems where there is a potential for exposure either in manufacturing, use or demilitarization. If the potential for release to the environment is present, environmental fate, transport, and ecotoxicity is evaluated. Typically, Toxicity Assessments can serve as a technical foundation resource for other broader evaluations (e.g. PESHE) that are requirements for the program.

8.3.3 ESOH Data Requirements

As previously mentioned, few specific data requirements are addressed for many military programs. Because of previous experience, approaches to the acquisition of chemical/physical property and toxicity data have been developed. Many follow and/or build on simple green chemistry procedures and are phased with the level of investment dedicated to program research [4]. Some studies, because of the relative effort involved, are best handled in the acquisition stages.

8.3.3.1 Approaches

Some regulatory agencies have specific requirements for chemicals based on annual production volume and other criteria (e.g. US Environmental Protection Agency [USEPA], European Union (EU) [e.g. Registration, Evaluation, Authorisation and Restriction of Chemicals, REACH]). Many approaches used in green chemistry applications begin with an iterative interaction between developer and the ESOH professional, where ESOH criteria (e.g. toxicology, chemical/physical properties) are considered equivalent alongside other performance criteria. Initially, data that are commensurate with the level of investment devoted to the research program are collected. This typically means a high reliance on screening assays (with high false-positive rates, low false-negatives) that are inexpensive and high in uncertainty; however, they provide value for relative comparisons. Additional information is provided as follows.

Phased Approach Generally, phased approaches are recommended for obtaining data for use in ESOH assessment. As previously mentioned, specific ESOH data are collected alongside other performance criteria and considered as such when making decisions to continue research and development and, ultimately, implementation. Two examples that are specific to military applications [4, 5] exist, although others have been developed for applications such as pharmaceuticals and other industrial materials [6]. Logically, specific data are required at specific levels of research or technology research levels (TRLs; Figure 8.1). As suggested, this process begins early in conceptualization and builds on subsequent information through acquisition

Each test or assay provides specific data that can be initially used in categorization (e.g. Globally Harmonized System categories) [7]. Data can also be used in relative comparisons (e.g. when each substance is tested simultaneously) for interpretations of data collected from many *in vitro* tests when target tissue concentrations are not known.

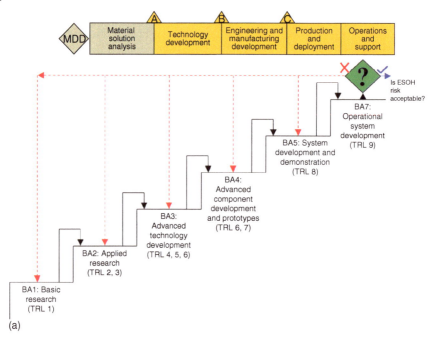

(a)

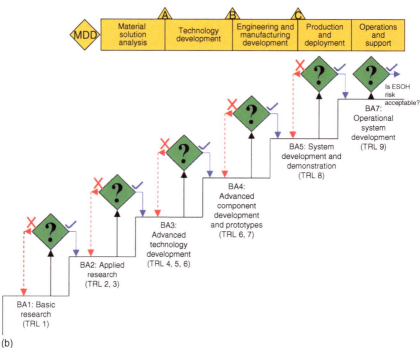

(b)

Figure 8.1 Unacceptable (a) and optimal (b) conceptual approaches to obtaining and assessing environmental, safety and occupational health information with increasing technology research levels within research and acquisition programs.

Optimally, each subsequent data collection event (e.g. bioassay) would build on knowledge collected from previous events, providing less uncertainty at each step. As the materials are tested and determined to be efficacious in their intended application, more detailed toxicity data are collected to better ensure the health of the warfighter and worker and for sustainable use in the environment.

Evolving Science and New Tools An integral part of tiered or phased approaches initially assumes a relatively low level of effort and cost that increases as the level of success and investment increases in the development of a new substance to be used in a weapons system. Inversely proportional to the low initial level of investment is the level of uncertainty of the information that is obtained. Conceptually, as greater success is realized in research and development, greater resources are devoted to understanding the toxicology of new substances and formulations and the uncertainty associated with potential adverse ESOH outcomes is proportionally reduced.

The science and tools now available in the area of toxicology, fate, and transport has broadened and is increasing. The advent of high-throughput screening and the US National Academy of Sciences publication of toxicity testing in the twenty-first century has become a guideline force in advancing the use of new biomedical science in toxicity evaluation [8]. However, many of these tools are lacking in terms of refinement – to aid in the understanding of variation in results and the uncertainty associated with the outcome. In some cases, additional research is needed to ascertain the level of precision and accuracy in test results.

However, some of these new bioassays and tools clearly have current applicability in the initial stages when acceptable levels of uncertainty are relatively high. Computational (*in silico*) chemistry models that only require knowledge of chemical structure to provide toxicity and chemical/physical property estimates that can be used in predicting fate and transport can be used during conception stages. Read-across methods, where chemicals with toxicity have similar chemical structures, can be used to infer relative toxicity of new molecules that are structurally similar. New *in vitro* methods can be used in early synthesis stages alongside similar substances (or replacements) for relative comparisons. The use of appropriate positive and negative controls is informative and provides confidence in the results. Although *in vitro* results often neglect important exposure factors that define target dose to tissue, the results can be used in a relative comparison to replacements. Note that it is not yet advisable to use these methods as alternatives to others before production and use, as many are still exploratory and a greater level of certainty may be needed. However, information gained can be used to build on subsequent stages and save time and resources on more focused *in vivo* tests when greater certainty is required. See further chapters in this series for more specific information.

Research vs Testing Generally, good research develops more questions than it addresses. Developing ESOH information requires the opposite. Tests to be used to gather ESOH data (e.g. toxicology tests) necessitates that resources used to obtain such information be such that they can be used to make risk-based decisions. Therefore, care must be taken to collect such data with a clear understanding how various data outcomes would be used *a priori* to make a decision. If it is determined that these data would not allow for an ESOH-based decision to be made, then the test should not be conducted.

In addition, it is likewise important that qualified institutions conduct such tests and that subject matter experts in the field be consulted beforehand. Often, regulatory agencies focus on positive (i.e. statistically significant) results and negative data rarely have weight when any positive relationships exist. Simply put, positive results (even false-positive) are rarely discounted and are perpetuated; it is exceedingly difficult to prove a negative.

Integration of Flow Charts   With advancing science occurring at a fast pace, new and emerging methods may be used to replace older ones. In addition, society is increasingly demanding reductions in the use of *in vivo* studies (laboratory animals) for ethical as well as financial reasons. Therefore, a question-based matrix is best used to help address ESOH concerns. Such a paradigm can be used throughout the process from initial to final development stages. Time to complete these requirements will vary depending on the outcomes of the data collected at earlier stages, pathways of exposure, and complexity of the system.

It is important to note that ESOH concerns are broad and complex. These questions are not new and few comprehensive approaches in developing a question-based key exist. Toxicity depends on exposure potential and exposure depends on system design, manufacture, engineering controls, and nature of environmental release (e.g. wastewater discharge, incomplete high-order detonation, low-order detonation, or dud frequency, etc.). The value in such an approach is that it allows the ESOH professional to use whatever tools that are available to address the likelihood of the impact. The drawback is that it can be complex and the right assay or tool would be needed to address the specific question. The level of uncertainty must be addressed at each stage.

Addressing exposure can initially be done through an analysis of the chemical's physical properties (Figure 8.2). For example, if the vapour pressure is relatively high, then exposure can be assumed to occur via the inhalation pathways and inhalation toxicity data would be needed. If the new compound has high water solubility, then environmental releases would be assumed to reach groundwater or move through surface/sheet water run-off. The former would require oral toxicity data for human health applications and the latter would require aquatic ecotoxicity information. An example is provided in Figure 8.3.

Reproduction/Developmental Effects   Recently, many countries have changed or are changing policies associated with the role of women in military service. In many circumstances, women of child-bearing age are now on the front lines of military service where first-hand exposures to obscurants, combustion products from weapon systems, and exposures to other materials (e.g. smoke mixtures, pyrolysis products from fire extinguishing agents, etc.) can have the potential to affect the normal development of unborn children at a period potentially before these warfighters realize they are pregnant. This is particularly important in exposure scenarios like submarine environments that are functionally similar to

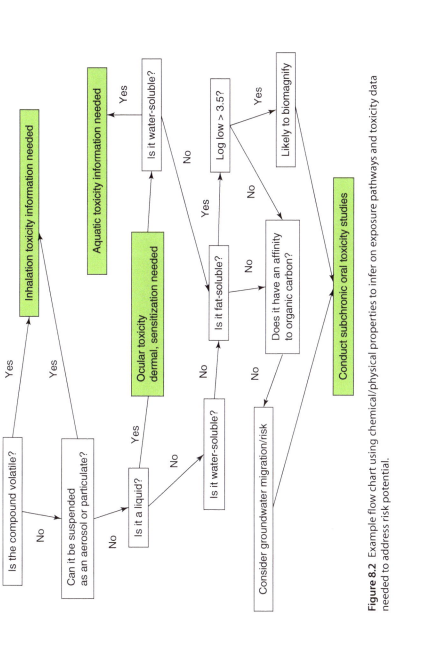

Figure 8.2 Example flow chart using chemical/physical properties to infer on exposure pathways and toxicity data needed to address risk potential.

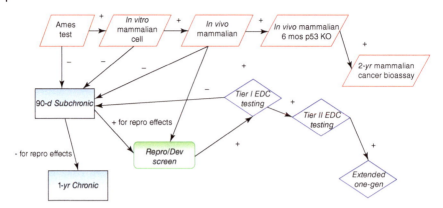

Figure 8.3 Example flow chart with appropriate bioassay to be used in a phased manner to address the potential for genotoxicity. Boxes with blue background are used as part of other flow charts to understand potential for non-genotoxic events.

confined spaces. Relatively, very few chemicals have developmental toxicity data to enable the determination of risk from exposure. Often, adverse developmental outcomes originate from substances that have a genotoxic mode of action – that is, exposure, at early life stages affect genetic structure or translation and cause developmental abnormalities. However, environmental factors can also cause adverse developmental outcomes.

Adverse reproductive effects (i.e. substances that adversely affect the normal operation of the reproduction system) can occur from a variety of mechanisms. Compounds that act to impede or accentuate the endocrine system or those that can be directly toxic to germ or nurturing cells (e.g. sertoli cells) can reduce fertility.

Assessing the potential of a compound to cause adverse reproductive or developmental effects is relatively complex and requires integration into other existing data plans. Typically, it is reasonable to investigate for potential reproductive effects when there is a potential for exposure, when there is evidence of endocrine disruption, and/or when there is evidence that the reproductive organs are affected from repeated exposures using animals (e.g. a subchronic rodent bioassay).

In the past, failure to address the impact of reproductive effects of substances that are released into the environment has caused marked changes in the ecosystem and has led to much legislation (e.g. DDT effects on predatory birds, PCBs, etc.). If effects are observed either from histological change to reproductive organs as part of a repeated dose *in vivo* study or if there are *in vitro* data that suggest endocrine affinity of the new molecule, then a phased approach to investigating reproductive effects is needed. Examples of tiered approaches for testing the potential for endocrine disruption can be found at https://www.epa.gov/endocrine-disruption.

Understanding pathway-specific toxicity potential for systemic genotoxic effects is important in predicting some specific toxic outcomes. Compounds that cause changes to the genome can conceptually lead to the development of cancer for the developing foetus (developmental effects). Hence, other flow charts can be developed to help understand the importance of this pathway of toxicity (Figure 8.3). If the results of these tests are positive, it is also advisable that an *in vivo* testing paradigm that is designed to consider development effects be used.

8.4 Current and Evolving Regulatory Interests

The EU and the United States have initiated recent legislation to more comprehensively evaluate the ESOH effects from use for many new compounds conceptually planned for production. Specific military exemptions do exist; however, current sustainable use while training, safe manufacturing, and safe use by the warfighter suggests the collection and assessment of ESOH data to be a worthwhile endeavour with a high potential for a return on investment.

Many jurisdictions and corporations involved in chemical research and production realize that starting early asking these questions has a conceptual and practical basis. However, they note that the abundance of chemicals in commercial production exceeds the capabilities to conduct thorough and complete vertebrate bioassays. Therefore, many new *in silico* and *in vitro* tools are being developed. Regulatory agencies are devoting resources into research to explore ways to extrapolate such data to human dose. *In vitro*–*in vivo* extrapolation (IVIVE) is an area that is growing in the research community; however, much work still needs to be done to have solid applications in the determination of safe levels of exposure for workers and soldiers. Regardless, such data are helpful in a weight-of-evidence context in using techniques such as read-across in helping to focus *in vivo* studies and reduce the use of animals in toxicity studies.

Clearly, not every new substance must be tested, but should be evaluated if there is significant exposure potential and probability. Probability for exposure can be determined and ranked where those with the highest rank can be evaluated first. Iterative discussions with system developers can be most useful in designing our problems and finding successful options.

8.5 Case Studies and Cost Analysis

The process from development of new chemical structure to implementation to a weapon system is complicated and requires many qualification processes and steps. Few examples currently exist as this process is relatively new. However, some can be discussed as follows.

8.5.1 M116, 117, 118 Simulators

The 'whistle, bang, flash' simulators are used in training exercises to emulate combat simulations. The fuel for these simulators previously contained potassium perchlorate, which was implicated in contributing significantly to groundwater contamination. The perchlorate anion was found not to be acutely toxic; however, it was subsequently found to interfere with the sodium-iodide symporter in the thyroid, affecting thyroid hormone production and thyroid cell growth and potentially leading to cancer [1]. The US Army Environmental Quality Technology Program, Pollution Prevention Pillar devoted resources into the development of less toxic alternatives. Subsequently, systems that used a black powder alternative (potassium, sulfur, sodium nitrate) that was less expensive and maintained the same level of expected performance were developed.

8.5.2 M-18 Violet Smoke

M-18 coloured smoke grenades are designed as signalling tools, are activated and thrown, and hence exposure was considered to be minimal. However, subsequent use in urban situations has led to using these grenades as obscurants, where exposure to the warfighter was greatly increased with adverse health consequences occurring as a result. Much of the toxicity was occurring from combustion of the fuel (hydrogen sulfide production); therefore, the fuel was reformulated replacing the sulfur with sugar. This change created a change in the internal thermal dynamics where dyes needed to be reformulated and the grenade re-engineered. Initial attempts using Solvent Violet 9 found acute toxicity (mortality) in rats at the limit dose from a single four-hour exposure. Equivalent performance results at no additional costs were obtained using a combination of solvent red and solvent blue dyes, which we subsequently found not to result in acute toxicity at the limit dose. This iterative exchange of information between developer and ESOH professional led to the development of a less toxic alternative with no increase in cost. Further development is under way.

8.5.3 Cost and Time Considerations

Cost and time to conduct studies to obtain ESOH information has always been an issue. Researchers and program managers often complain of the limited resources available and that conceptually these additional data requirements risk reducing available funding for research, development, and/or production, regardless of the great potential for return on investment. To date, funding devoted to ESOH data needs compared to those for research have been typically less than 10% of the total program. Still, researchers and program managers need cost and schedule projections to plan accordingly.

Table 8.1 provides cost and time schedules based on enquiries conducted with government laboratories and contract research organizations between 2015 and 2016. Although these values are likely to change, they provide a relative measure for comparison purposes.

Inherent in the collection of these data are resolutions to ultimate and proximate questions. This means these data can be used in a relative way to see if the new technology will be less toxic than previous systems or some data can be used to determine how much exposure would occur and what the risk would be. Examples of the latter would include using inhalation toxicity data to develop an occupational exposure level for industrial hygiene purposes – a requirement

Table 8.1 Chemical/physical properties of ESOH use.

Property/attribute	Utility
Molecular mass (MW)	Determine dermal flux, understand excretion rates, and pathways
Water solubility (mg/l)	Environmental fate and transport, exposure potential (e.g. gut absorption potential)
Fat solubility (octanol/water partition coefficient; log K_{ow})	Potential for gastrointestinal absorption and bioaccumulation/magnification between trophic levels
Vapour pressure (Torr)	Potential for inhalation exposures; environmental ½ life
Affinity to organic carbon (log K_{oc})	Fate and transport; soil sorption, potential to reach ground water from release
Henry's law coefficient	Environmental ½ life in surface water
Boiling point	Inhalation potential; environmental persistence
Melting point/ionization potential	Fate and transport

Source: ASTM [4].

Table 8.2 Relative cost estimates of selected toxicology tests needed for assessment of chemical and material hazards (c. Jan. 2013).

	Total			
Toxicology tests	Average estimation costs (k$)	SD	Average estimation duration (days)	SD
Acute oral	4	2.3	74	12
Sub-acute oral (14 days)	109	24.4	36	8
Sub-acute oral (28 days)	135	14.6	61	32
Sub-chronic oral (90 days)	207	30.3	90	0
Chronic oral (1 year)	580	—	685	—
Chronic oral (2 years)	1200	—	1200	—
Acute inhalation (limit)	11	3.3	9	8
Sub-acute inhalation (28 days)	77	—	90	—
Acute dermal toxicity (limit)	4	2.6	32	4
Primary skin irritation	3	1.8	22	11
Primary eye irritation	3	1.8	30	7
Skin sensitization	13	5.5	33	1
Neutral red uptake (cytotoxicity)	6	5.3	12	13
Ames test	8	3.6	29	8
Mouse micronucleus assay	23	9.4	24	25
Chinese hamster ovary (CHO) test	21	3.2	24	25
Hepatic COMET assay	58	70	na	na
Mouse lymphoma assay	28	11.38	60	—

before manufacturing can occur. Another example would include using aquatic toxicity data to help determine whether release of manufacturing wastewater would be appropriate and legal to the general wastewater stream. These data can also be used in sophisticated environmental soil models to ensure training and testing ranges can be used sustainably (Table 8.2).

8.6 Summary

This chapter reviews the logic of and background for the collection of ESOH data to be used as additional performance criteria in the assessment of new molecules for weapons systems and platforms. It provides the benefits and costs of tests and how they may be used in decision-making. It puts forth a logic and method to integrate data into a preventive medicine/public health paradigm that will help support the development of safe, sustainable energetics for use in the theatre, and at training ranges worldwide. As scientific advancements are made, this paradigm is flexible to include such advancements and provides the rationale for validation and refinement of new technologies as they are developed.

Acknowledgements

I thank Erik Hangeland and Kimberley Watts of the US Army Research Development and Engineering Command, the US Army Environmental Quality Technology, Pollution Prevention Program and Dr. Robin Nissan of the Strategic Environmental Research and Development Program (SERDP) for supporting this work. Thanks go out also to Bill Ruppert for contributions to figures.

References

1 Mattie, D.R., Strawson, J., and Zhao, J. (2006). Perchlorate toxicity and Risk Assessment. In: *Environmental Occurrence, Interactions and Treatment* (ed. B. Gu and J.D. Coates), 169–196. New York, NY: Springer. ISBN: 978-0-387-31113-5.
2 Agency for Toxic Substances Disease Registry (ATSDR) (2012). Toxicological profile for RDX. U.S. Department of Health and Human Services, Public Health Service, Atlanta, GA.
3 Agency for Toxic Substances Disease Registry (ATSDR) (1995). Toxicological profile for 2,4,6-trinitrotoluene (TNT). U.S. Department of Health and Human Services, Public Health Service, Atlanta, GA.
4 American Society for Testing and Materials E 2552 (ASTM) (2016). Standard guide for assessing the environmental and human health impacts of new compounds for military use. In: *Section 11, Water and Environmental Technology, Biological Effects and Environmental Fate, Biotechnology*, vol. 11.06. Conshohocken, PA: ASTM International.

5 The Technical Cooperative Program (TTCP) (2014). Assessing the Potential Environmental and Human Health Consequences of Energetic Materials: A Phased Approach. *TTCP Tech. Rep. No. TR-WPN-TP04-15-2014*. Brochu, S. (CP lead), Hawari, J., Monteil-Rivera, F., Sunahara, G., Williams, L.R., Johnson, M.S. (US POC), Simini, M., Kuperman, R.G., Eck, W.S., Checkai, R.T., Cumming, A.S. (UK POC), Doust, E., Provatas, A. (AU POC), 1–62.
6 Manley, J.B. (2015). The five Ws of pharmaceutical green chemistry. In: *Green Chemistry Strategies for Drug Discovery* (ed. E.A. Peterson and J.B. Manley), 1–12. Cambridge, UK: Royal Society of Chemistry.
7 United Nations (2003). *A Guide to the Globally Harmonized System of Classification and Labelling of Chemicals*, 1–90. New York, NY: United Nations Press https://www.osha.gov/dsg/hazcom/ghsguideoct05.pdf.
8 National Research Council (NRC) (2007). *Toxicity Testing in the 21st Century: A Vision and A Strategy*. Washington, DC: National Academic Press.

9

Overview of REACH Regulation and Its Implications for the Military Sector

Carlos Ferreira, Fausto Freire, and José Ribeiro

University of Coimbra, ADAI-LAETA, Department of Mechanical Engineering, Portugal

9.1 Introduction

Environmental and safety legislation has been a driver to make industry develop new approaches to mitigate, or at least decrease, the environmental impacts of their products. Ammunition production activities are not an exception to this. A clear example of legislative pressure is the REACH (Registration, Evaluation, Authorisation and Restriction of Chemicals) regulation that aims to enhance the protection of human health and ecosystems as well as improve the competitiveness of the European chemical industry [1, 2].

With REACH, the European authorities aim to encourage, and in some circumstances to guarantee, that substances reported as harmful are replaced by available and viable (economically and technically) alternatives with lower hazard [2]. Harmful substances, named as substances of very high concern (SVHC), are classified accordingly with criteria outlined in Article 57 of REACH, which identify the physico-chemical properties and toxicity effects considered important for the characterization of its hazard to human health and ecosystems.

REACH regulation can pose a significant concern for European defence capabilities and the associated industrial production as they affect the full life cycle of military systems (design, procurement, production, use, and disposal). Substances such as dibutyl phthalate (DBT) (a plasticizer used in gun propellants) already need an authorization to be produced or imported into Europe, and others (e.g. 1,3,5-hexahydro-1,3,5-trinitrate (RDX); 2,4-dinitrotoluene; di-isobutyl-phthalate) are also at risk and may also be restricted. Any restrictions in the production, importation, and use of these substances can seriously impact the operability of the defence industry, as it is difficult to comply with REACH regulation and at the same time present alternatives which maintain the required performance level [3].

These restrictions for certain chemical substances are driving both research laboratories and industry to search for or to develop safer alternatives. However, the alternatives that can be proposed may also present problems with hazard effects on

Energetic Materials and Munitions: Life Cycle Management, Environmental Impact and Demilitarization,
First Edition. Edited by Adam S. Cumming and Mark S. Johnson.
© 2019 Wiley-VCH Verlag GmbH & Co. KGaA. Published 2019 by Wiley-VCH Verlag GmbH & Co. KGaA.

human health and ecosystems, and eventually produce other types of environmental impacts. Due to the importance of REACH regulation for the munition industry, this regulation needs to be comprehensively examined to understand how the classification of hazardous substances is carried out, and to ascertain if the toxicological hazard classification system for chemical substances is missing some important aspects to properly strengthen the reliability of the identification of SVHC. Therefore, this chapter intends to present an overview of the important aspects of the REACH regulation, focusing on the authorisation process with the aim of presenting some recommendations to improve the classification of hazard substances.

9.2 Regulation for Hazard Substances

This section presents a historic summary of the legislation concerning the hazard identification of chemicals in Europe, which drove the development of the REACH regulation. An overview of the principal aspects of the REACH regulation is presented, and the limitations concerning the classification of harmful substances are also discussed.

9.2.1 Overview of Previous Legislation Concerning Hazard Substances in the European Union

Before the implementation of the REACH regulation, chemical substances were classified for toxicity according to a legal imposition, or a self-classification was made by the producers [4]. Warhurst [5] described in detail the regulatory system before REACH, stating that the system could be divided into four types of regulation:

i) *Council Directive 67/548/EEC* was the first regulation published in Europe, which created a standardized system for classification, packaging, and labelling of dangerous substances. Since its publication, this directive has been revised many times, and in 2008 it was replaced by the Regulation (EC) No. 1272/2008 – Classification, labelling and packaging of substances and mixtures (which is a complement to REACH regulation);
ii) *Directive 76/769/EEC*: Restrictions on the marketing and use of certain dangerous substances and preparations – provided restrictions for specific uses of chemicals. This regulation only restricted the use of substance in the market, instead of completely banning their use;
iii) *Directive 79/831/EEC*: New chemicals regulation – required that new substances, produced in quantities higher than 10 kg/year after 19 September 1981, need to go through some type of assessment before being introduced in the market. This regulation is more proactive than the restriction on use, which only applies if it is shown that substances already in the market are dangerous;
iv) *Regulation 793/93*: Evaluation and control of the risks of existing substances – applied to the substances already in the market. This regulation can also be seen as proactive as it aims to look and evaluate chemicals of concern, instead of responding to an alarm.

The previous legislation had some drawbacks associated with the lack of information related to toxicity and the uses of the existing chemicals from which arise a great difficulty of the regulator in assessing the risks. These drawbacks resulted in what was considered insufficient protection of human health and ecosystems [5] and led to the creation of the REACH regulation. Meanwhile, awareness of toxicity exposure and risk increased amongst activists, the media, and the general population. Therefore, the pressure on governments to act and protect the population from chemical hazards also increased. Another driver that helped create the REACH regulation was the gradual development of a legal agenda for risk management, which includes the precautionary principle[1] [7].

9.2.2 Overview of REACH Regulation

REACH is a European Union regulation concerning the Registration, Evaluation and Authorisation of Chemicals. REACH places on industry the responsibility of assessing and managing the risks that chemicals can present for human health and the environment. The principal aim of REACH is to enhance the protection of human health and ecosystems as well to as improve the competitiveness of the European chemical industry [1, 2]. Another important objective of the REACH regulation is to encourage, and in some circumstances to guarantee, that substances reported as harmful are eventually replaced by available and viable (economically and technically) alternatives with lower hazard [2].

The REACH regulation includes three main processes – Registration, Evaluation, and Authorisation of chemicals. Registration is intended to register all chemicals manufactured or imported at a level of more than 1 tonne/year in Europe. In contrast with previous regulation, no distinction is made between new substances or substances already in the market [8]. The information required to perform such registration is related to the uses of the substances and lists physico-chemical and toxicity properties which need to be disclosed to the European Chemical Agency (ECHA). As the tonnage produced or imported per year increases, the information required to perform the registration also increases [9]. It is important to note that for chemicals manufactured or imported at smaller amounts than 1 tonne/year, the REACH regulation body has no legal authority over them.

A new characteristic of the REACH regulation is that industry needs to develop a Chemical Safety Assessment (CSA) for all substances produced in quantities of 10 tonnes or more per year. The CSA process identifies and describes the situations in which a substance can be considered safe in its production and use [10]. This process includes three main steps: the hazard assessment, an exposure assessment, and risk characterization. In accord with ECHA guidance [10], the hazard assessment aims to "identify the hazards of the substance, assess their potential effects on human health and the environment, and determine, where

---

[1] The precautionary principle calls upon decision-makers to adopt measures when scientific information is insufficient, inconclusive, or uncertain about the potential effects on human health hazard and on the environment, but the risks are high [6].

possible, the threshold levels for exposure considered as safe (the so-called no-effect levels)". For that purpose, in this step, industry needs to collect all available and important information about the chemical, which includes data about its properties, the production process, types of uses, and potential emissions and exposures. Should the information be considered insufficient to satisfy the REACH conditions, more information must be provided. The CSA ends in this first step if a substance is identified as not hazardous or not classified as persistent, bioaccumulative, or toxic (PBT), or very persistent, very bioaccumulative (vPvB). However, if it is observed that a chemical meets at least one of the criteria mentioned before, the other two steps of the CSA need to be performed.

The second step – exposure assessment – is based on the uses of the substances, in which the exposure for humans and the environment to a dose or concentration of a certain chemical [10] is estimated (or measured). In this step, the exposure scenarios that determine the critical levels of exposure of humans and the environment for the production and use of a substance is defined. These scenarios take into account all identified uses and life stages of a substance [10].

Risk characterization is the final step, in which the levels of exposure are compared with the limits defined for each effect. REACH considers that a risk is controlled if the exposure levels are under the limits defined for toxic effects on humans and the environment. The risk characterization is undertaken with a qualitative or semi-quantitative approach where the effects do not have a published limit. In these cases, the risk is considered controlled if the exposures are mitigated or avoided [10].

For the cases in which the risks are not controlled, the CSA needs to be completed with additional information that includes more data about the properties of the substance and the presentation of restrictions in the production or use, or through obtaining more accurate estimations of the exposure. This process ends when the risks are considered under control. However, a substance can be recommended not to be used if the risks are considered uncontrolled for a specific use (and no more iterations to different exposure scenarios are either possible or economically viable). In the end of this process, the different steps of the CSA need to be documented in the Chemical Safety Report (CSR).

Evaluation is the step in which the government authority evaluates the information provided in the registration. The information is checked to ensure compliance with REACH, assessing the danger a substance can pose to human health or the environment [9]. The evaluation can lead to additional testing, recommendations, or restrictions [8].

In the authorisation processes, the SVHC are identified accordingly with criteria outlined in Article 57 of the REACH regulation [2], which include substances that are:

i) carcinogenic,
ii) mutagenic,
iii) toxic for reproduction
iv) PBT
v) vPvB, or
vi) substances that pose an equivalent level of concern to those listed in the previous points.

Annex XIII of the REACH regulation establishes the thresholds for the values of some properties for the classification of the substances as PBT or vPvB. Exceeding only one of the thresholds can be sufficient to classify a substance as PBT or vPvB [2]. It is important to mention that PBT and vPvB substances are identified by comparing the values of some of the properties of the substances with quantitative thresholds; whilst the other criteria outlined in Article 57 are based on a qualitative evaluation of scientific evidence of those effects on human health or the environment. Table 9.1 lists the thresholds established for the identification of PBT or vPvB chemicals [2, 11].

Potentially harmful substances are nominated to a candidate list and prioritized to be restricted or banned [9]. For these substances, an authorization for its use is required, which can be obtained if one of two requirements are met [8], which are:

i) demonstrate that the use of the chemical is controlled and safer; or
ii) that there is a socioeconomic need for its use.

Even if the aforementioned elements are guaranteed, the Authorisation is given only after analysing the opinions of the Committee for Risk Assessment and the Committee for Socioeconomic Analysis. However, it must be noted that for SVHC, the authorization for its use shall not be applied.

Some substances can be only restricted or limited in their use, as stated in Article 68, which can be proposed by a member state and afterwards approved by the EU governing organizations. Those substances shall not be produced, imported, or used without complying with the conditions of the agreed restriction.

9.2.3 Discussion of REACH Regulation

The REACH regulation was an important advance for the identification of hazardous chemicals by providing harmonized registration and authorisation processes for chemicals produced or imported in Europe. In addition to existing

Table 9.1 Limits of the physico-chemical and toxicity properties established in Annex XIII of REACH regulation.

Properties	PBT	vPvB
$t_{1/2}$ marine water	>60 days	>60 days
$t_{1/2}$ freshwater	>40 days	>60 days
$t_{1/2}$ soil	>120 days	>180 days
$t_{1/2}$ marine sediment	>180 days	>180 days
$t_{1/2}$ freshwater sediment	>120 days	>180 days
BCF	2000	5000
NOEC (or EC10)	<0.01 mg/l	<0.01 mg/l
LD_{50} oral	<200 mg/kg	<25 mg/kg
LC_{50} inhalation	<10 mg/l	<2 mg/l

approaches, it considers that existing and novel chemicals have the same level of relevance and that the potential hazard of both needs to be evaluated. The aim of not only identifying hazardous substances but also hastening the procurement of safer alternatives also helps the protection of human health and ecosystems.

Despite the evident advantages that REACH regulation has given to the evaluation of hazard chemicals, there are some concerns regarding the quantitative identification of PBT (or vPvB) chemicals carried out by comparing the values of some properties of the substances with thresholds for the same properties from the REACH regulation.[2] This classification is established in Annex XIII; and from what is proposed, the following principles can be drawn and the following problems/questions arise:

i) Hazard is essentially dependent on the individual value of a limited number of the substance properties. The thresholds established refer to the properties persistence (marine water, freshwater, soil, and marine sediment), bioaccumulation factor (BCF), and toxicity. However, it is important to understand why only these properties were considered whilst other physico-chemical properties (persistence in air, water solubility, vapour pressure), that can influence the behaviour of a substance in the environment, and consequently its hazard, were not considered. Petry et al. [7] show some apprehension in prioritizing substances based solely on certain inherent physico-chemical properties, which they believe can bias the classification.

ii) The regulatory thresholds established for the different substance properties are equivalent. Nevertheless, the equivalence of the values selected for the thresholds was not verified. The values considered for the thresholds for each property can have a different weight in assessing the potential hazard, so a certain property may have a greater importance on that evaluation than others considered for the PBT identification. The equivalence of the values selected for the thresholds should be verified by evaluating if the toxicological consequence of the physico-chemical and toxicity properties separately, results in the same hazard.

iii) The combinatory effect of the different substance properties is not relevant. The identification of PBT chemicals is carried out by comparing the individual values of each property with its respective threshold, so the influence of the combination of different properties to the potential hazard is considered irrelevant. The toxicological impact of a substance results from its behaviour in the environment, the exposure to that substance, and from its toxicity. This is not determined by the value of the substance properties individually, but by the combination of all properties and also of the environment characteristics such as average temperature, type of soil, wind direction, rain rate, etc. The influence of the combinatory effect of all the substance properties and the environment characteristics cannot be neglected in the evaluation of the potential hazard of a chemical substance. Therefore, the correctness of this

---

[2] These issues are discussed in more detail in the article by Ferreira, C., Ribeiro, J., Freire, F. A hazard classification system based on incorporation of REACH regulation thresholds in the USEtox method (submitted in Journal Cleaner Production).

approach should be evaluated to understanding the importance of the combinatory effect in the hazard. The inclusion of the combinatory effect of the properties is also important when evaluating safer alternatives to ensure that all potential hazard consequences are accounted for.

Blainey et al. [12] also states that substances that do not exceed the classification thresholds can still pose a potential hazard to ecosystems and human health, due to the potential importance of the combinatorial effect of different physico-chemical and toxicity properties on their impact. For instance, a substance with values of properties that do not exceed any of the thresholds considered in the PBT classification can still pose a significant hazard to human health and ecosystems, since the combination of all properties with the environment characteristics can originate a high exposure that can increase the risk.

iv) The unrestricted production and use of substances with values of properties exceeding the limits established in Annex XIII is unacceptable. The selection of the values for thresholds also needed some attention and clarification on why they were considered an acceptable reference for human hazard, and substances exceeding that limit need to be replaced by suitable alternatives. Since the REACH regulation became active, specific frameworks or methodologies were proposed in the literature to help in the identification or in the design of new, alternative substances complying with the existing regulations. Some of those methodologies perform a preliminary chemical PBT screening based on molecular configuration [13] while others are more detailed and include in that evaluation aspects beyond the regulatory issues. Askham et al. [1], for example, developed a tool that goes beyond the REACH classification system and considers environmental and economic performance indicators. It converted them into a single score. Bruinen de Bruin et al. [14] proposed a framework to set up an environmental impact assessment to be included into the socio-economic analysis of the REACH regulation, and Giubilato et al. [15] proposed a framework with the employment of different occupational exposure models to perform a comparative risk assessment for different occupational exposure scenarios of an industrial production process.

A different approach, which follows the principle of REACH regulation, focusing only on hazard aspects in order to perform screening or priority-setting PBT substances, is now proposed. This can be useful as decision support for REACH or other policies. The approach is based on the employment of toxicological characterization factors (CFs; as calculated by the USEtox method) to evaluate the toxicological hazard. This appears to be a better way to identify the potential hazard substances than a classification based on a simple comparison of the values of some of its properties related to PBT with the respective thresholds. The calculation of the toxicological CFs includes more properties of the substances than the PBT classification system and also the combinatory effect of those properties and the characteristics of the environment. Another important difference between these two classification systems is that the PBT classification system is based on a qualitative evaluation of the hazard to humans, and so does not provide information of how much lower the potential hazard of alternative substances is when compared with one of the restricted substances, while the employment of toxicological CFs allows a quantitative evaluation of that difference.

To demonstrate the feasibility of this approach, a comparative assessment of the toxicological CFs associated with DBT and four alternatives is described to determine the substance with lower potential hazard. DBT is a substance used in gun propellant production that, due to its toxicity for reproduction and aquatic toxicity, is included in the list of substances for potential substitution. Four alternatives to DBT are considered:

i) diisononylphthalate (CAS 28553-12-0);
ii) dioctylsebacate (CAS 122-62-3);
iii) dioctyl terephthalate (CAS 6422-86-2); and
iv) oxydiethane-2,1-diyl dibenzoate (CAS 120-55-8).

Table 9.2 presents the values of the physico-chemical properties for DBT and the four alternatives. The values are taken from the ECHA database, which is provided on the ECHA website (http://echa.europa.eu/pt/home) for consultation. The only exception is the half-lives that were determined with the EPI Suite™. The red values indicate that the substances exceed the PBT limits presented in Annex XIII of REACH regulation (or other thresholds that were also considered on the basis of the values presented in the literature).

It is difficult to determine the substance with lower (or higher) hazard based solely on the comparison of values of the properties presented with the respective thresholds. Each one of the alternatives exceed one, or more, of the thresholds established in Annex XIII or taken from the literature, which means that in the future these alternatives can also be considered for some type of restriction in use. One approach that can be used to overcome this problem is to calculate the toxicological CFs associated with these substances. This will provide a quantitative manner to compare the alternative substances with the restricted ones. Therefore, the next step is to calculate the toxicological characterization factors with the USEtox method to assess which substance presents lower potential toxicological impacts on human health and ecosystems.

The toxicological characterization factors for DBT and the four alternatives were calculated with the USEtox method, considering the properties presented in Table 9.2 (more information regarding how the calculation is made is available in Chapter 7 of this book). The toxicological CFs were calculated for the impact categories of ecotoxicity and human toxicity (for this latter, the acute effects were the only ones considered for the calculation as none of these substances have carcinogenic data available), and for six different emission compartments (urban air, rural air, freshwater, seawater, natural soil, agricultural soil) to cover all the potential media in which an emission can occur. The toxicological impacts are always calculated by the USEtox method in excel (or a software package), which is provided by the USEtox creators, for new substances or the ones that already exist in the USEtox database.

Figure 9.1 shows the toxicological characterization factors for DBT and the four alternatives for the impact category human toxicity, considering six emission scenarios; whilst Figure 9.2 shows the toxicological characterization factors

Table 9.2 Physico-chemical and toxicity properties associated with dibutyl phthalate and the four alternatives selected.

Properties	Dibutyl phthalate	Diisononyl phthalate	Dioctyl sebacate	Dioctyl terephthalate	Oxydiethane-2,1-diyl dibenzoate	PBT limits
$t_{1/2}$ air (day)	1.15	0.46	0.30	0.48	0.56	>2[a]
$t_{1/2}$ water (day)	8.66	37.5	15	15	15	>40
$t_{1/2}$ soil (day)	17.3	75	30	30	30	>120
$t_{1/2}$ sediment (day)	77.91	337.5	135	135	135	>120
BCF (l/kg)	No data	3.0	6.6	393	60	>2000
Log K_{ow} (–)	4.57	9.7	10.08	7.81	3.2	>3.5
K_{oc} (l/kg)	1.4	6.0	397100	117489	1500	
Sol$_{25}$ (mg/l)	10.0	0.0006	2.0	0.0004	38.3	>1[a]
k_H (Pa (m^3/mol))	0.124	41.4	No data	No data	No data	—
P_{vap} (Pa)	0.013	0.00006	0.0002	0.001	0.000018	—
LD$_{50}$ oral (mg/kg)	6279	10000	4560	5000	3535	<200
LC$_{50}$ inhalation (mg/l)	15.68	4.4	3.2	No data	200	<10 mg/l
EC$_{10}$ (mg/l)[b]	0.1; 0.1; 0.2; 10	24.5; 101; 88	352	0.28; 0.76; 0.86; 10	3.3; 100	<0.01
Properties of concern	Toxic to reproduction	No hazards classified	No hazards classified	Suspected carcinogen	No hazards classified	Carcinogenic, mutagenic, toxic to reproduction

a) The thresholds for these properties are obtained from the literature – $t_{1/2}$ air from Scheringer et al. [16] and Sol$_{25}$ from Vighi and Calmari [17].
b) The values presented are related to the aquatic toxicity for fish, invertebrates, algae, and microorganisms (by that order).

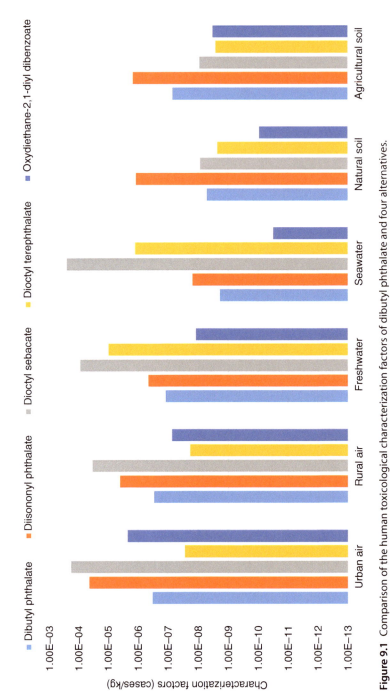

Figure 9.1 Comparison of the human toxicological characterization factors of dibutyl phthalate and four alternatives.

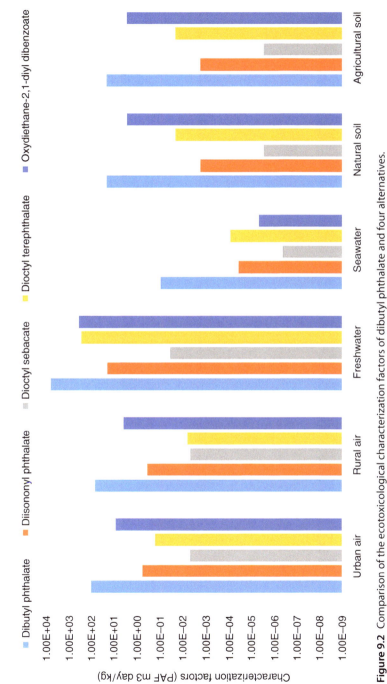

Figure 9.2 Comparison of the ecotoxicological characterization factors of dibutyl phthalate and four alternatives.

for the same substances and emission compartments but for the impact category ecotoxicity. It is observed that the results differ with the emission compartment considered or the toxicological effect (human toxicity vs ecotoxicity). For human toxicity, the substance with lower toxicological impact for an emission into air (urban and continental) and agricultural soil is dioctyl terephthalate and for an emission into water (freshwater and seawater) and natural soil is oxydiethane-2,1-diyl dibenzoate. For ecotoxicity, dioctyl sebacate gives the lower toxicological impact for all emission compartments.

The fundamental aim of this analysis is to show that it is easier to comprehend which substance(s) pose a lower potential toxicological impact on human health and on ecosystems, than by analysing each substance's property individually using comparison with thresholds from regulation or legislation. Another fundamental difference between these two approaches is that the REACH PBT classification system is based on a qualitative evaluation of the human hazard, and so does not provide information on how much lower the potential hazard of alternative substances is when compared with one of the restricted substances, while the employment of toxicological CFs allows a quantitative evaluation of that difference.

The results show that even using the toxicological CFs as a base for comparison, different conclusions on which substance is the appropriate alternative can be made as the results vary for different emission scenarios and toxicity effects. This trade-off can be overcome if the CFs of a reference substance, defining an acceptable level of the potential human hazard, is used for comparison and screening. As it is difficult to find a set of real substances to be used as reference, a hypothetical substance can be defined for that purpose. This hypothetical substance should have as many properties as those necessary for the calculation of the toxicological CFs using the USEtox method, being the value of those properties defined on the basis of the thresholds that the REACH regulation used for the PBT classification. The others are arbitrarily set to the most frequent value of those properties for the most common organic substances. This approach has some limitations as the analysis is still a hazard assessment in which the risk or toxicological impact is not accounted, and the identification of the PBT substances is made on the exceedance of thresholds (the advantage of using the new approach is that the thresholds for the toxicological CFs permits considering more properties than the ones used by the REACH regulation and also consider the combinatory effect of those properties and environment characteristics).

This approach follows the principle of REACH regulation focusing only on hazard aspects to perform screening or priority setting for PBT substances, which can be useful as decision support for REACH or other policies. The identification of more than one substance that complies with the regulatory thresholds, based on a comparison with a reference, is of great importance as other environmental or performance characteristics can change the selection of the safer substance. This occurs principally when a shift of impacts is observed, for instance, the selection of a substance that presents lower toxicological impacts, but either has a higher impact in the production phase or is required at a higher quantity to perform the same function as the conventional products. The preliminary analysis of the toxicological characterization factors with the inclusion of the toxicological thresholds also supports decision-makers in presenting reasons and advantages for the selection of certain alternatives, or to present justifications for the absence of appropriate alternatives to substances in Annex XIV of the REACH regulation and thus for their continued use.

9.3 Conclusions

This chapter described the main implications of the REACH regulation to the military systems, since REACH regulation intensifies the management and use of hazardous substances which will affect the availability of energetic materials and other materials for munitions. The section of alternative materials for future application is also constrained by this regulation. Therefore, it is important to analyse the impact of the REACH regulation to ensure that military industry provides the munitions and equipment for European countries that meet the specific requirements for protection of human health and ecosystems. For that purpose, a summary of the legislation concerning the hazard identification of chemicals in Europe that originated the development of the REACH regulation was given. An overview of the principal aspects of the REACH regulation was outlined and its impact on munitions and their environmental management considered. The limitations concerning the classification of harmful substances were discussed. A practical example of those limitations was shown with the assessment of alternatives to DBT using the REACH classification system. The use of toxicological CFs in the classification and screening of safer alternatives for substance banned (or restricted) by REACH, as an example of an approach to solve/mitigate those limitations, was illustrated and shown to be effective for munition constituents as well as for general materials.

While REACH is an EU initiative and regulation, it does affect other nations both through the need to understand the regulation in order to trade with the EU and as an imperative to develop similar or related regulations of their own. It, therefore, has a wider impact than just within the EU.

References

1 Askham, C., Gade, A.L., and Hanssen, O.J. (2012). Combining REACH, environmental and economic performance indicators for strategic sustainable product development. *J. Cleaner Prod.* 35: 71–78.
2 European Commission (2006). Regulation (EC) No. 1907/2006 of the European Parliament and of the council of 18 December 2006 concerning the Registration, Evaluation, Authorisation and Restriction of Chemicals (REACH), establishing a European Chemicals Agency, amending Directive 1999/45/EC and repealing Council Regulation (EEC) No. 793/93 and Commission Regulation (EC) No. 1488/94 as well as Council Directive 76/769/EEC and Commission Directives 91/155/EEC, 93/67/EEC, 93/105/EC and 2000/21/EC.
3 EDA (2016). Study on the impact of REACH and CLP European chemical regulations on the defence sector. Executive Summary of Final Rep., REACHLAW Ltd., Brussels, Belgium.

4 Oltmanns, J., Bunke, D., Jenseit, W., and Heidorn, C. (2014). The impact of REACH on classification for human health hazards. *Regul. Toxicol. Pharm.* 70: 474–481.
5 Warhurst, A.M. (2006). Assessing and managing the hazards and risks of chemicals in the real world – the role of the EU's REACH proposal in future regulation of chemicals. *Environ. Int.* 32: 1033–1042.
6 Bourguignon, B. (2015). *The Precautionary Principle – Definitions, Applications and Governance.* European Parliamentary Research Service. ISBN: 978-92-823-8480-0.
7 Petry, P., Knowles, R., and Meads, R. (2006). An analysis of the proposed REACH regulation. *Regul. Toxicol. Pharm.* 44: 24–32.
8 Hansen, S.F., Carlsen, L., and Tickner, J.A. (2007). Chemicals regulation and precaution: does REACH really incorporate the precautionary principle. *Environ. Sci. Policy* 10: 395–404.
9 Cohen, A.K. (2011). The implementation of REACH: initial perspectives from government, industry, and civil society. *Int. J. Occup. Environ. Health* 17: 57–62.
10 ECHA (2009). *Guidance in a Nutshell: Chemical Safety Assessment.* European Chemicals Agency, ECHA-09-B-15-EN. ISBN: 978-92-95035-11-9.
11 European Commission (2008). Regulation (EC) No. 1272/2008 of the European Parliament and of the council of 16 December 2008 on classification, labelling and packaging of substances and mixtures, amending and repealing Directives 67/548/EEC and 1999/45/EC, and amending Regulation (EC) No. 1907/2006.
12 Blainey, M., Avila, C., and Zandt, P. (2010). Review of REACH annex IV – establishing the minimum risk of a substance based on its intrinsic properties. *Regul. Toxicol. Pharm.* 56: 111–120.
13 Gramatica, P., Cassani, S., and Sangion, A. (2015). PBT assessment and prioritization by PBT index and consensus modeling: comparison of screening results from structural models. *Environ. Int.* 77: 25–34.
14 Bruinen de Bruin, Y., Peijnenburg, W., Vermeire, T. et al. (2015). A tiered approach for environmental impact assessment of chemicals and their alternatives within the context of socio-economic analyses. *J. Cleaner Prod.* 108: 955–964.
15 Giubilato, E., Pizzol, L., Scanferla, P. et al. (2016). Comparative occupational risk assessment to support the substitution of substances of very high concern: alternatives assessment for diarsenic trioxide in Murano artistic glass production. *J. Cleaner Prod.* 139: 384–395.
16 Scheringer, M., MacLeod, M., Matthies, M., Klasmeier, J. (2006) Persistence criteria in the REACH legislation: critical evaluation and recommendations, Final Rep., Safety and Environmental Technology Group, Zurich.
17 Vighi, M. and Calamari, D. (1993). Prediction of the environmental fate of chemicals. *Ann. Ist. Super Sanità* 29 (2): 209–223.

10

Development and Integration of Environmental, Safety, and Occupational Health Information

Mark S. Johnson

U.S. Army Public Health Center, Toxicology Directorate, 8252 Blackhawk Road, Aberdeen Proving Ground, MD, 21010-5403, USA

10.1 Introduction

The use of energetic substances (explosives, propellants, and pyrotechnics) has been prevalent since at least the nineteenth century and continues. Initially, the indirect costs of manufacture and use were considered insignificant compared with primary mission goals. With increasing consideration of the health of our warfighters, workers, and the environment, understanding the health-related impacts of the manufacturing, use, and demilitarization of these substances has been a priority. Contamination of soil and water by explosives is a serious environmental problem that is costing the world economy hundreds of millions of dollars to manage and to remediate. In addition, occupational health (industrial hygiene) and environmental releases (e.g. from effluent discharges) during manufacturing require some of the same toxicology data to assess safety to workers and to provide the means for manufacturing. To begin production without a sound understanding of environmental safety and occupation health (ESOH) consequences can result in wasted resources, adverse health effects to exposed personnel, and/or cessation of training and testing activities that are necessary to maintain military readiness and comply with public health legislation requirements.

Pathways of environmental exposures are varied. Releases can occur during manufacturing from intentional and unintentional decisions (e.g. wastewater discharge and accidents, respectively). Environmental releases have the potential to contaminate valuable groundwater resources and migrate off ranges. Contaminated water can be a pathway for human, livestock, and irrigation exposures. Exposures to biota can occur from soil releases that can also result in surface run-off and migrate off ranges. The US Government Accountability Office projects that it could cost more than $121 billion to remediate ranges [1]. Many international jurisdictions have regulations that require the consideration of ESOH in the development, manufacturing, use, and demilitarization of these substances and systems; however, few provide any guidance on how to consider these attributes or what data are needed in this assessment or evaluation.

Recently, there have been several efforts in the United States to develop a phased approach to the development of an ESOH data requirement. Two reports are final and other two are draft (personal communication), yet all are relatively consistent with each other and describe specific data needs that are associated several general levels of compound development and subsequent use [2, 3]. The steps and processes are subsequently described herein.[1] Other international efforts in the development of drugs for human health and veterinary applications follow paradigms similar to those of many proprietary green chemistry approaches adopted by industry.

10.2 Phased Approach to a Toxicology Data Requirement

The subsequent material provides a rational, tiered approach to development of specific data that can be used to address hazards from environmental use and releases and from occupational exposures. It begins with the initial research on a novel substance with a relatively small level of effort for ESOH data collection and proposes that decision-makers consider these criteria alongside other performance metrics in the assessment for continued effort and resources. Scientific management decision points occur between each stage, with the level of uncertainty in data collection narrowing as the technology progresses. Subsequent data collection builds on previous efforts (Table 10.1).

Research, development, testing, and evaluation (RDT&E) is identified by the stages of conception through demonstration. Acquisition of new technology typically includes manufacturing and use and demilitarization. As these terms are general ones, specific data requirements should be considered by each jurisdiction when making policy decisions requiring data collection. This is important in meeting cost and schedule milestones.

10.3 Research, Development, Testing, and Evaluation

10.3.1 Conception

Energetic compound development typically begins with the *in silico* exploration of functional moieties that have the properties necessary for a desired performance. Often these molecules are notional and may not currently exist. These computer-generated models can predict chemical/physical properties of interest to both the developer and the health professional. These properties are described here and include their ESOH relevance.

---

[1] *Note to readers*: Although the concepts presented are logical and straightforward, users are encouraged to engage professionals in toxicology to ensure completeness in the testing paradigm and in interpreting the results. This is critical in meeting cost and schedule requirements.

Table 10.1 Data and *in silico* models available to predict chemical properties and toxicology from Reference [3].

Data/endpoint	Databases and estimation programs	URL
Basic physical–chemical properties		
Melting point	MPBPWIN™	http://www.epa.gov/oppt/exposure/pubs/episuite.htm
Boiling point	ACD/Labs percepta predictors	http://www.chemspider.com/
Water solubility	WSKOWWIN™; WATERNT™ ACD/Labs percepta predictors	http://www.epa.gov/oppt/exposure/pubs/episuite.htmhttp://www.chemspider.com/
Volatilization, vapour pressure	WVOLWIN™ ACD/Labs percepta predictors	http://www.epa.gov/oppt/exposure/pubs/episuite.htmhttp://www.chemspider.com/
Henry's constant	HENRYWIN™	http://www.epa.gov/oppt/exposure/pubs/episuite.htm
pKa and dissociation constant(s)	ACD/Labs percepta predictors	http://www.chemspider.com/
Density/specific gravity	ACD/Labs percepta predictors	http://www.chemspider.com/
K_{ow}	KOWWIN™ ACD/Labs percepta predictors	http://www.epa.gov/oppt/exposure/pubs/episuite.htmhttp://www.chemspider.com/
Fate and transport		
Transport		
Fugacity	LEV3EPI™	http://www.epa.gov/oppt/exposure/pubs/episuite.htm
Exposure	E-FAST	
K_{oc}	KOCWIN™ ACD/Labs percepta predictors	http://www.epa.gov/oppt/exposure/pubs/episuite.htmhttp://www.chemspider.com/
Bioaccumulation	BCFBAF™ ACD/Labs percepta predictors	http://www.epa.gov/oppt/exposure/pubs/episuite.htmhttp://www.chemspider.com/
Degradation		
Hydrolysis	HYDROWIN™	http://www.epa.gov/oppt/exposure/pubs/episuite.htm
Biodegradation	BIOWIN™; BioHCwin	http://www.epa.gov/oppt/exposure/pubs/episuite.htm
Exposure	E-FAST	
Toxicology		
Mammalian		

(Continued)

Table 10.1 (Continued)

Acute and chronic oral toxicity	TOPKAT®	
Acute and chronic inhalation toxicity	TOPKAT	
Acute and chronic dermal toxicity	TOPKAT	
Acute and chronic ocular toxicity	TOPKAT	
Development and reproduction	TOPKAT	
Mutagenesis/ carcinogenesis	TOPKAT	
Non-mammalian		
Aquatic/terrestrial plant toxicity	ECOTOX release 4.0	http://cfpub.epa.gov/ecotox/
Aquatic/terrestrial invertebrate toxicity	ECOTOX release 4.0	http://cfpub.epa.gov/ecotox/

In silico estimates of toxicity may also be made from chemical structure. Quantitative structural activity relationship (QSAR) models use regression analyses from comparison to chemical structural similarities with other compounds with known toxicity data sets. Estimates are often provided with qualitative indicators of confidence associated with each based on the rigor and variation in the database and relationship to the representativeness of the chemical structure within the database. QSAR models are used in pharmacology and many exist. Specific models are available as freeware for property estimation (quantitative structural property relationship, QSPR), for estimating laboratory animal toxicity and for aquatic plant and animal toxicity [4]. These are also often referred to as computational chemistry approaches.

Note that reliable estimates of toxicity are often lacking for energetics as functional moieties are often poorly represented in the database. Moreover, as functionally what makes them unique (i.e. as strong oxidizers, rarely organic) suggests that, as a class, energetics do not follow property relationships as traditional organics. Energetics tend not to be lipophilic and often not water-soluble and frequently form crystals.

This means they often have a low affinity to organic carbon fractions of the soil and leach to groundwater. High water solubility of salts only increases this probability.

Where poor *in silico* predictions occur, read-across methods may be employed. It only involves understanding the fate, transport, and effects of compounds that have similar molecular structures and infer probability of transfer and effects. This approach requires professional judgement and is greatly enhanced where there is molecular data for specific moieties to help make evidence-based decisions regarding the fate, transport, and toxicity of new molecules with a class. Classes of energetics include nitroaromatics (benzene ring with adjoining nitro groups; e.g. 2,4,6-trinitrotoluene, tetryl, 1,3,5-trinitrobenzene, 2,4-dinitotoluene, 2,6-dinitroanisole), nitrate esters (nitroglycerine, pentaerythritol tetranitrate (PETN), diethylene glycol dinitrate (DETN)), and nitramines (hexahydro-1,3,5-triazine, RDX; 1,3,5,7-tetranitro-1,3,5,7-tetrazocane, HMX).

Releases of nitroaromatics to wet soils generally result in the reduction of the nitro groups to amines and subsequent binding of the amine group to humic fractions of the soil, making them unlikely to transport in the environment [5]. Effects from exposures to nitroaromatics typically result in effects similar to those of nitrates where sublethal effects include anaemia. These substances are also more likely to transverse the skin, making dermal exposure important.

Nitramines tend not to adhere to soil or be very water soluble and have crystal-forming properties. Generally, they move through the vadose zone and can reach groundwater. Absorption from oral exposures can be rapid (e.g. RDX) or very minimal in monogastric species (e.g. HMX).

Nitrate esters typically release nitric oxide in the bloodstream following exposure and cause vasodilation. This is the fundamental mechanism of action for nitroglycerin pharmacological benefit for patients with angina, and PETN was also tested for therapeutic use early in its discovery. Toxicity includes the refractory effect of vasoconstriction following activation that has been reported to be the cause of headache in munition factory workers. Environmental releases of nitrate esters can be persistent.

10.3.2 Synthesis

Following conception, small amounts of investigated material are synthesized. This is often relegated to small, desk-top operations where only gram quantities are made. At this stage, specific data are experimentally collected (see Table 10.2).

The science of toxicology is rapidly expanding where new high-throughput *in vitro* techniques may be used. Each is typically very specific and can be used if specific pathways for toxicity are expected on the basis of read-across or other data. They should be used only when reliable conclusion can be made from the results, i.e. when sufficient interlaboratory verification and validation test have been completed. They should *not* be used when there is an insufficient database of results and information does not exist regarding probability for false-negative or false-positive results (Type I and Type II errors). When a positive result is found, it may be difficult to understand even when more reliable data are subsequently generated. Relative comparisons of results for these assays can, however, be useful when paired with appropriate positive and negative controls.

It should be noted that data should be collected using standardized procedures to ensure quality assurance and control. Toxicology tests should be conducted according to the principles in good laboratory practices to be acceptable by some regulatory agencies. Professional judgement is an integral part of this process and should not be undertaken by individuals or agencies not qualified in interpreting data in a public health context.

Table 10.2 Data and methods needed at the synthesis stage.[a]

Data-endpoint	References[a]
Basic physical–chemical properties	
Individual soluble constituents	
Melting point	OECD 102 (1995a)
Boiling point	OECD 103 (1995b)
Water solubility	OECD 105 (1995c)
Vapour pressure – Henry's constant	OECD 104 (2006a)
Density/specific gravity	OECD 109 (2012a)
Viscosity	OECD 114 (2012b)
Dissociation constants in water	OECD 112 (1981a)
K_{oc}	OECD 121 (2001a)
K_{ow}	OECD 123 (2006b)
Photolysis	OECD 316 (2008c), USEPA 712-C-08-013 (2008c)
Hydrolysis	OECD 111 (2004a), USEPA 712-C-08-012 (2008b)
Ready biodegradability or inherent biodegradability	OECD 301 (1992c), 310 (2006f), OECD 302C (2009c)
Toxic properties of all constituents	
Cytotoxicity	
Microtox (*Vibrio fischeri*)	Choi and Meier (2001)
Mammalian cell line – neutral red exclusion	OECD 432 (2004f)
Ligand binding affinity (e.g. steroid receptors)	OECD 455 (2012f)
Cancer	
Mutagenicity – Ames assay	OECD 471 (1997b)
Mammalian cell gene mutation	OECD 476 (1997e)

a) All references cited in [3].

10.3.3 Testing/Demonstration

During this stage, scaling-up processes have been established for synthesis of new compounds, and formulations are being explored for specific applications. Typically, this results in larger quantities of material where exposure is markedly increased for those working and investigating this technology. Here, focused *in vivo* work is expected to understand the toxicokinetics as well as the toxicodynamic aspects of exposure. In addition, since specific applications are under way, more information regarding manufacturing, use, and disposal as well as propensity for exposure can be described (Table 10.3).

Table 10.3 Data requirements for testing and demonstration.

Data-endpoint	References[a]
Mammalian toxicity	
Non-cancer	
Acute oral toxicity	ASTM 1163 (2010), OECD 401 (1987a), OECD 420 (2002d), OECD 423 (2002e), OECD 425 (2008e), USEPA EPA 712-C-02-189 (2002a), USEPA 712-C-02-190 (2002b)
Subacute (28-day) oral toxicity	ASTM 1163 (2010), OECD 407 (2008d), USEPA 712-C-00-366 (2000)
Potential for metabolism	
Acute inhalation toxicity	ASTM 1291-99R03 (2003b), OECD 403 (2009d), OECD436 (2009g), USEPA 712–C–98–193 (1998b)
Acute dermal toxicity	OECD 402 (1987b), USEPA 712–C–98–192 (1998a)
Skin sensitization	OECD 406 (1992e), OECD 429 (2010d), USEPA 712–C–03–197 (2003)
Skin irritation/corrosion	OECD 439 (2010e), OECD 404 (2002c), USEPA 712–C–98–196 (1998d)
Eye irritation/corrosion	ASTM 1055-99R03 (2003a), OECD 405 (2012e), USEPA 712–C–98–195 (1998c)
Cancer	
Genotoxicity/clastogenicity, *in vitro*	OECD 476 (1997e), OECD 482 (1986b), OECD 479 (1986a), OECD 473 (1997c), OECD 483 (1997f), OECD 486 (1997g)
Genotoxicity/clastogenicity, *in vivo*	OECD 474 (1997d)
Fate, transport, and effects	
Individual soluble constituents	
Stability and transport in soil	ASTM E1197-87R04 (2012a), OECD 312 (2004e), 304A (1981c)
Bioconcentration and bioaccumulation	ASTM E1676 (2012c), Walsh et al. (2012), NRC (2003), Van Gestel and Ma (1988), Jager et al. (2005), OECD 315 (2008b), OECD 317 (2010b)
Formulations	
STANAG – closed vessels	NATO (1997)
Detonation residues	Pennington et al. (2006)

a) All references cited in [3].

10.3.4 Acquisition

After the technology has been developed and has satisfied functional requirements for use, chemical formulations and accompanying technologies are transferred to system production. The following sections provide guidelines for meeting standard ESOH criteria to satisfy health and often regulatory requirements, although these may vary depending on jurisdiction. As before, the user is encouraged to seek professional assistance in requesting studies, discussing range of potential outcomes and potential risks, and interpreting data to meet cost and schedule projections.

Interpretation of toxicity data may be needed to develop functional tools for production. For example, results from subchronic rodent bioassays can be used to develop occupational exposure levels (OELs) for industrial hygiene purposes. Data from other assays can be used in a weight-of-evidence approach to develop values and potentially skin notations. Physical and chemical property information can be used to develop personal protection equipment. Aquatic toxicity data collected using standardized methods approved by regulatory agencies can be used to develop wastewater discharge permits. Limits on exposures to sensitive populations can be made, if necessary (e.g. if exposure could cause developmental effects, then exposure to women of child-bearing age would be scrutinized).

10.3.5 Engineering and Manufacturing

Technology development is largely complete by this stage and plans are being made to refine processes and implement production. A clear understanding of exposure potential should now be known, and data are needed to ensure worker, warfighter, and environmental health can be managed. Methods are needed to conduct industrial hygiene at ammunition facilities, OELs need to be developed, environmental permits may be required to handle hazardous waste, and acceptable range loads should be known. Understanding the environmental consequences of use as intended should be clear, where environmental loads from high-order and low-order detonations are characterized. Other jurisdictions may have other specific requirements to ensure safety to the warfighter, worker, and the environment. Having these suggested data sets will assist in providing information to address these concerns.

The minimum toxicity data set to develop safe levels of exposure for workers or the general population (OEL) is the 90-day subchronic assay conducted in rodents. These employ rigorous methods that explore other endpoints for toxicity. If effects that suggest the reproductive system is a target are found, then endocrine disruption potential must be evaluated. These assays are also conducted in a phased manner and begin with *in vitro* and limited *in vivo* protocols.

As with reproductive or developmental effects, data from subsequent stages that suggest genotoxicity, other tests which are more reliable, yet intensive are indicated.

Substances used in munitions that are water-soluble and/or have a high probability for reaching groundwater could become an attribute that could halt training and cause health threats in migrating groundwater plumes that may migrate off installations and be used as drinking water sources for public consumption. In addition, substances that can be found in the soils following use could represent a chronic hazard for exposed personnel. In these cases, chronic (long-term) toxicity tests are needed to develop acceptable daily doses for exposed individuals. When such data are not available, regulatory entities may choose to employ an uncertainty factor to ensure the safety of exposed civilians. In some cases, the conduct of such chronic tests can result in a significant benefit to the manufacturer and the military, particularly when extended exposures are not likely to result in adverse effects occurring at a lower exposure level. Failure to meet regulatory or health-based criteria may cause expensive remedial options, harm to public health, and result in cessation of training and testing activities.

Manufacturing needs may also require cleaning/maintenance of the facility where discharge of material into the wastewater streams is needed (e.g. laundering of cotton overalls). In these cases, specific data are needed to determine if these materials may be digested and attenuated through wastewater treatment facilities and not result in toxic effects to aquatic biota. Many jurisdictions use standardized protocols to determine effects on aquatic systems; some of which can be found in Table 10.4. Some protocols are chemistry-focused, where inoculum from a wastewater treatment facility is used under laboratory conditions to ascertain whether the planned effluents can be broken down if released.

Programme managers are also encouraged to understand the nature and extent of other exposure pathways to additional personnel. Naturally, exposures to warfighters must be determined not to be harmful to enable the successful completion of the mission. This includes a broad understanding of all potentially in the theatre of operations. Examples include individuals engaged in explosive ordnance disposal (EOD), training, vehicle maintenance, and civilians (Table 10.4).

10.3.6 Demilitarization

That munitions, weapons systems, and platforms will be decommissioned, with contents often recycled is an emerging theme within the Armed Services. "Design for Demilitarization (DFD)" concepts have been developed and used, having their origins from past environmental issues that caused health-based cessation of training and use activities and/or large-scale remedial operations. Developers are encouraged to engineer systems that allow for easy disassembly and recycling of materials. Best practices for DFD are provided in the available manual [6, 7].

Past practices that involve open burning/open detonation (OB/OD) are regulated in many jurisdictions and, while practical in theatre operations, have become impractical or prohibited in testing operations in many countries. Generally, data from the past decade have been consistently trending downwards where fewer and fewer regulatory authorities in the United States are permitting such operations. Efforts must be made to understand the magnitude and impact of demilitarization and to consider such in munitions design wherever possible.

10.4 Other Data Requirements

10.4.1 Environmental

10.4.1.1 Fate and Transport

Methods that provide information regarding exposure is useful when planning and scheduling toxicity studies. Methods that can provide data on fate and transport may suggest that groundwater contamination is unlikely and that chronic data may not be needed. In addition, these data may also be useful in determining

Table 10.4 Tests for engineering and manufacturing.

Data-endpoint	References[a]
Mammalian toxicity	
Non-cancer	
Subchronic (90-day) mammalian oral toxicity	OECD 412 (2009e)
Subchronic (90-day) mammalian inhalation toxicity	OECD 413 (2009f)
Endocrine disruption, oral *in vivo*	OECD 440 (2007b)
Reproductive/fertility effects, oral	USEPA (2009)
Reproductive/developmental toxicity, repeated oral dose	OECD 415 (1983), OECD 416 (2001c), OECD 421 (1995e), OECD 42 2 (1996b)
Advanced toxicokinetics	OECD 417 (2010c)
Neurotoxicity, delayed effects	OECD 418 (1995d), OECD 424 (1997a)
Developmental neurotoxicity, oral dose	OECD 426 (2007a)
Developmental neurotoxicity, inhalation exposure	OECD 426 (2007a)
Cancer	
Carcinogenicity bioassay oral	OECD 451 (2009h)
Fate, transport, and effects	
Formulation	
Thermal stability/stability in air	OECD 113 (1981b)
Propellant residues	Ampleman et al. (2011), Diaz et al. (2012), Walsh et al. (2012, 2011)
Explosive residues	Pennington et al. (2006)
Individual constituents	
Adsorption/desorption	OECD 106 (2000a), 312 (2004e)
Biodegradability in soil (various)	OECD 307 (2002a)
Aerobic mineralization in surface water	OECD 309 (2004d)
Biodegradability in sediment (various)	OECD 308 (2002b)
Biodegradability in seawater	OECD 306 (1992d)
Bioconcentration and bioaccumulation	ASTM 1676 (2012c)
Abiotic and biotic breakdown products	Hawari and Halasz (2002), Hawari et al. (2000), Zhao et al. (2004), Sunahara et al. (2009), Halasz and Hawari (2011), Paquet et al. (2011)
Water/wastewater treatment	
Biodegradability in wastewater	OECD 314 (2008a)
Anaerobic biodegradability in activated sludge	OECD 311 (2010a)
Wastewater treatability	
Aerobic sewage treatment	OECD 303 (2001b)

a) All references cited in [3].

range loading that would not result in the likelihood of off-site migration or adverse effects to nearby populations. See Table 10.4 for a list of data considerations for fate and transport.

10.4.1.2 Ecotoxicity

Many jurisdictions have regulations that protect wildlife species at ranges against harm – that may be considered from exposure to products from munitions. Threatened and endangered species and/or birds are often protected. In these cases, toxicity data for non-mammalian species may be useful in addressing these concerns. In addition, specific aquatic toxicity information can be valuable to manufacturing (e.g. wastewater release permits) as well as for decision-making/clean-up operations for unexpected releases. It is important that standardized, regulatory-endorsed methods be consulted before testing is begun.

10.4.1.3 Field Monitoring

When fate and transport data are incomplete or uncertain, it is advisable to engage in real-time monitoring efforts to determine whether use could result in a substantive release. Perimeter groundwater wells may be used to periodically test for materials. Effectiveness of blow-in-place operations may be determined from soil constituent residues. These actions, done in a phased manner, may be useful in the identification of environmental issues before exposures occur.

10.4.1.4 Disposal

As in DFD, recycling material is recommended and in many cases are cost-effective. However, use as intended and blow-in-place may result in significant sources of environmental release. The data collected in previous levels should provide information that will help define the most effective manner of disposal when material exists.

10.4.1.5 Occupational – Industrial Hygiene

As mentioned previously, a minimum data requirement exists for the development of an OEL. It is advisable that systems managers plan early, as these tests typically take nearly a year and cost ~$280 000 to complete. Dermal toxicity data are also needed to determine extent of personal protective equipment (PPE) needed. Following test completion, study results are often interpreted by experts (often through peer review) to develop an OEL. Without such data, extraneous PPE may be recommended to protect the worker from any residual uncertainty when toxicity databases are incomplete. Few quality criteria exist for substance data sheets (SDSs) and care is indicated using this information without additional confirmation. PPE are typically more difficult to remove after more health data are collected; therefore, it is advisable to collect useful toxicity data beforehand.

In addition to OELs, industrial hygiene personnel need accepted analytical chemistry methods that are repeatable and include specific quality control and assurances. Specific tools may be needed given the relative toxicity.

Consider exposure potential for all engaged personnel, which may be exposure through the manufacturing and use of the material. EOD personnel may be exposed to a higher concentration than others, and levels of PPE may be minimal. Qualified industrial hygienists may be helpful in recognizing exposure potential and mitigating this to reduce potential for effects. Consider the potential for exposure to those engaged in maintenance also.

10.4.2 Regulatory

10.4.2.1 Toxic Substance Control Act

Recently, the Frank R. Lautenberg Chemical Safety Act for the twenty-first century became law and effectively is a revision of the Toxic Substances Control Act first enacted in 1976. The new law differs in the following significant ways: (i) requires that manufacturers prove safety, (ii) requires the US Environmental Protection Agency (USEPA) to now prioritize and mandate testing, if necessary, to more than 80 000 chemicals in commercial production. Many of these were grandfathered in from the previous legislation, and (iii) use *in vitro* methods and weight of evidence procedures where and when they are scientifically justified to do so. Currently, the USEPA has ranked and begun the process on 20 substances. It is not known precisely how this new legislation will affect weapon systems development; however, a higher level of rigor in review and providing toxicity data are expected.

10.4.2.2 REACH

The Registration, Evaluation, Authorisation, and Restriction of Chemicals (REACH) is a European Union regulation that was enacted in 2007. The European Chemical Agency (ECHA) is the responding agency tasked with reviewing data for many substances. Predominantly, ECHA is focused on the use of *in vitro* data, although the use of *in vivo* data is recognized. Additional aspects on REACH are addressed in other sections and hence not repeated here.

10.4.3 Integrating Weight-of-Evidence into Decision-Making

Interpreting toxicology data into a risk assessment context is rarely straightforward, and often there is significant variation in how regulatory agencies value data points and how they are used to develop regulatory guidance or promulgated values. Thousands of toxicity data points may be developed, and understanding the relative importance of each is different from protective (screening) relative to predictive approaches. Using the same data for different subpopulations can also result in significant variation (e.g. OELs vs values intended for the general population). Discriminating between false-negative and false-positive results is also important, as is extrapolating results from laboratory animals to humans. Procedures exist to value, even semi-quantitate studies that can assist with the development of toxicology-based benchmarks [8–11]; however, the reader is encouraged to seek the assistance of qualified professionals in using these concepts in a risk assessment framework.

10.5 Concluding Remarks

Assessments of past practices have shown a reduction in unexpected events and increased probability to maintain cost and schedule requirements when weapons systems are designed considering ESOH criteria along with conventional performance metrics. A changing regulatory environment coupled with increasing encroachment activity at or near installations has heightened the need to understand the ESOH consequences of manufacturing, use, and demilitarization. Developing toxicity, fate, and transport data can be done in a phased manner that is commensurate and proportional to the level of investment devoted to the specific research, development, manufacturing, or use of the material (often <10%). Provided here are harmonized methods that can be used in these phases that will provide toxicology data that can be built on with subsequent phases and, therefore, at each stage reduce the uncertainty associated with further development and use. Employing a weight-of-evidence approach in assessment of these data will enable useful derivation of health-based toxicity benchmarks (i.e. safe levels) of exposure. Professional assistance is advised in the selection of studies and evaluation of data.

Acknowledgements

I thank Drs. William Eck, Lawrence Williams, Sonia Thiboutot, Guy Ampleman, and Sylvie Brochu who supported the development of the tables as part of a similar TTCP effort. This project was partially funded by the Strategic Environmental Research and Development Program (ER-2223) and the Army Environmental Quality Technology Program, Pollution Prevention through the US Army Research, Development, and Engineering Command (RDECOM), Environmental Technology Acquisition Program (Director, Eric Hangeland; Deputy Director, Kimberley Watts).

References

1 U.S. General Accounting Office (2004). DOD Operational Ranges: More Reliable Cleanup Cost Estimates and a Proactive Approach to Identifying Contamination are Needed. *GAO-04-601. Rep. to Congressional Requestors*, Washington, DC.
2 ASTM International ASTM E2552 (2016). Standard guide for assessing the environmental and human health impacts of new compounds for military use. In: *Water and Environmental Technology, Biological Effects and Environmental Fate, Biotechnology*. Conshohocken, PA: American Society for Testing and Materials.
3 The Technical Cooperative Program (TTCP) (2014). Phased Evaluation of Energetic Materials. *TR-WPN-TP04-15-2014. CP 4-42*. 1-62.
4 U.S. Environmental Protection Agency (USEPA) (2012). Assessment tools for the evaluation of risk (ASTER). Ecological structure activity relationships (ECOSAR) predictive Model. Risk assessment division. https://www.epa.gov/tsca-screening-tools/ecological-structure-activity-relationships-ecosar-predictive-model.
5 Johnson, M.S., Salice, C.J., Sample, B.E., and Robidoux, P.-Y. (2009). Bioconcentration, bioaccumulation, and biomagnification of nitroaromatic and nitrammine explosives in terrestrial systems. In: *Ecotoxicity of Explosives* (ed. G.I. Sunahara, G. Lotufo, R.G. Kuperman and J. Hawari), 227–252. Boca Raton, USA: CRC Press.

6 U.S. Department of Defense (2011). Manual: defense demilitarization. Program Administration No. 4160.28, Vol. 1, 1-34.
7 U.S. Department of Defense (2015). Defense materiel disposition manual: disposal guidance and procedures. DoD 4160.21-M-1, Vol. 1.
8 Klimisch, H.-J., Andreae, M., and Tillmann, U. (1997). A systematic approach for evaluating the quality of experimental toxicological and ecotoxicological data. *Reg. Toxicol. Pharmacol.* 25: 1–5.
9 Schneider, K., Schwartz, M., Burkholder, I. et al. (2009). ToxRTool, a new toll to assess the reliability of toxicological data. *Toxicol. Lett.* 189 (2): 138–144.
10 Wignall, J.A., Shapiro, A.J., Wright, F.A. et al. (2014). Standardizing benchmark dose calculations to improve science-based decisions in human health assessments. *Environ. Health Persp.* 122 (5): 499–505. https://doi.org/10.1289/ehp.1307539.
11 Gray, G.M., Baskin, S.I., Charnley, G. et al. (2001). The Annapolis accords on the use of toxicology in risk assessment and decision-making: an Annapolis center workshop report. *Toxicol. Mech. Methods* 11 (3): 225–231.

11

Research Priorities and the Future

Adam S. Cumming

School of Chemistry, University of Edinburgh, Joseph Black Building, The King's Buildings, David Brewster Road, Edinburgh, EH9 3FJ, UK

11.1 Introduction

Environmental legislative pressure may be assumed to continue in some form or other into the foreseeable future. The form that pressure will take will change, as will the priorities, and the overall importance placed on it may change, but continuing public pressure will provide impetus. Economic pressure is also likely to be maintained. Even if one nation might reduce the importance of this area, others will not, which will mean that freedom to ignore the issues will remain limited. What will change is the balance of importance and the move to addressing new problems or existing ones in a new way.

The other constraints on munitions, primarily vulnerability and life management and extension, also affect the problem as newer materials provide new challenges and present new problems. Existing methods are optimized for existing munitions and existing techniques will have a less than optimal effect on munitions designed for reduced vulnerability, etc.

The historical aspect must not be ignored. Dumps have been left after conflict or even after site closure and improved understanding raises the awareness of the possibility of toxicity or even explosion hazard. That some of these problems were assumed to be satisfactorily resolved at the time of creation is no consolation and may add to the problem. Sea dumping is one good example of this and, in general, handling materials in extreme conditions is a priority. This may include space as hazardous systems are in orbit….

One final factor that is significant and unlikely to change is the need to remove surplus materials from the market so that they are not available for misuse. While not exactly environmental science, it is a driver for investment as part of national security requirements.

When the need to understand the impact on the environment is added as a cost-saving measure there is clear reason for investment – it is always easier to plan ahead than to fix afterwards. However, emerging problems mean that the ability to fix these in a cost-effective manner is needed. In order to do this, there is a requirement to understand as much as necessary to deal with the problem. This may well be an 80% solution! Tools developed for preparatory assessments are now applicable and can provide fast assessment routes.

Therefore, a systematic approach to understanding and risk management is essential and the evolution of tools to do this will continue. Single-aspect studies remain important but will need to be linked to other aspects to deliver cost-effective and rapid solutions and management. Munitions will remain in service and in use, so the problem remains one of management.

Some of the areas of work have been reported in a recent issue of the journal, Propellants, Explosives and Pyrotechnics [1].

11.2 Greener Munitions

These have been covered earlier but efforts are certain to continue. The worked example of RIGHTTRAC (Revolutionary Insensitive, Green and Healthier Training Technology with Reduced Adverse Contamination) described in detail in Chapter 4 is a case where a complete and integrated approach has been taken and therefore represents a good approach to the problem. It is necessary to take such an integrated approach as this topic is a broad issue with many facets.

Changing public perception of environmental risk, which does in part drive new legislation, means that the environmental impact of munitions and their ingredients cannot be ignored. We require understanding of the problems if they are to be dealt with and questions answered, simply:

i) What is the impact of manufacturing processes as presently used and how may they be improved? Are there alternatives available or likely to become available?
ii) What is the effect of storage? How do different conditions affect the risks?
iii) What is the effect of use – on humans and on the environment?
iv) What are the toxicity effects in handling and use?
v) What are the effects on land – managing possible contamination?
vi) Are there disposal techniques available using safe methods?
vii) Can improved disposal methods be devised?

New processes can reduce manufacturing impact. Many processes were designed when there was less understanding of the biological effects and risks, and new approaches can be more efficient with reduced cost, while also producing better materials.

New materials can change the impact, and this needs to be understood, as was illustrated in Chapter 4.

However, 'greener' munitions are often discussed without reference to the real requirements for environmental impact and management. Such requirements are in addition to the prime purpose of the munition or propulsion system, which is to deliver the required performance. The need for a more systematic approach and assessment has prompted collaborative work worldwide. Both the requirements and the technology were assessed within a collaborative group and an example munition was studied in some detail, as described in Chapter 7. The requirements of design for disposal and the minimization of environmental impact were also reviewed with information from many places, and recommendations made for the future development of policies and tools for the environmental management of munition and rocket motor design.

There is a presumption that greener munitions means new ingredients, and certainly they have a role to play although only as part of a wider scheme. It must be recalled that energetic materials are intrinsically reactive and, therefore, as has been shown, have biological impact. Understanding this and managing it will continue to be an issue. New materials may prove to be safer, but may however introduce new problems or make existing ones worse. The work recently reported by Terreux and coworkers [2] gives one indication of how this might be managed. Prediction of bioactivity is important and will be a significant addition to the range of tools. Naturally any such tools must be validated and updated in the light of new understanding.

The US Strategic Environmental Research and Development Program (SERDP) is one of the organizations that are exploring the options for the future in Energetics. Their plan is shown below, courtesy of SERDP, and covers the broad spectrum of greener requirements.

The approach taken by SERDP offers one systematic route to an improved and improving methodology.

Environmentally Sustainable Manufacturing for Energetic Formulations is a recent program developed by SERDP and illustrates one of the critical factors.

Objective:
- Develop and mature novel technologies to reduce the environmental, safety, and occupational health impacts of the manufacture and load-assemble-pack (LAP) of energetic formulations.

Focus:
- Proposals should address high-use production or LAP methodologies, with a focus on current manufacturing processes.
- Technologies of interest include, but are not limited to, alternative solvents, solventless processing techniques, or novel mixing technologies for all energetic formulations, including primers and detonators, rocket and gun propellants, secondary explosives, and pyrotechnics.
- Details of this and other programs will be found on their website. https://www.serdp-estcp.org/.

11.3 Studies and Their Effect

The impact of the various studies in NATO and elsewhere has provided tools which are being employed where these issues are perhaps more pressing [3]. They have stimulated discussion and supported investment in other activities. Perhaps one of the most effective outcomes has been the development of an international network of experts. These links have produced knowledge which is being applied.

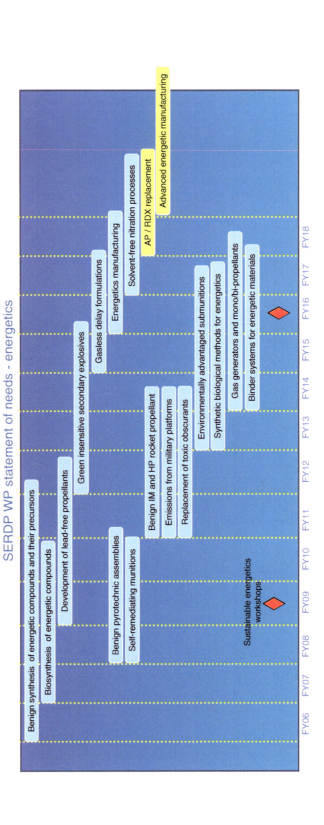

It has also helped identify questions and provide confidence in some of the answers. It has therefore also helped to identify gaps and look to the ways of filling them. They can be listed as follows:

Can we understand the basic properties of energetics well enough to perform the following?

Design new molecules and formulations;
Manufacture inexpensively and reproducibly;
Specify the component requirements, energetic or otherwise to meet the requirement in a cost-effective way;
Predict the behaviour of a material or a component, now and after ageing including environmental impact;
Operate a virtual munition design and testing laboratory with at least 80% effectiveness;
Be flexible enough to respond rapidly and cost effectively to new demands and new threats;
Integrate them effectively with other components making the best use of the properties of each to produce an optimal system or subsystem with the materials available.

This list is one that covers the whole of munition design and illustrates that environmental issues are and should be a part of the whole design and management approach.

11.4 The Problems and the Changing Requirements

Reading some of the literature may suggest that most of the problems have been solved, but this is far from the case. Changes in understanding can reveal problems.

One of the most important issues arising now is that of sea-dumped munitions. This has been covered in some detail in Chapter 6. Many munitions were dumped deliberately, as this was once considered a suitable method for disposal. However, these dumped munitions are causing problems as they often return and are washed up, or cause toxicity damage to the littoral area.

An example of this has occurred in the United Kingdom, where sea-dumped munitions – dumped in a trench in the continental shelf off Scotland's west coast – have been washed up on a tourist beach in Kintyre.

Wrecks and the debris of war will continue to cause problems of various kinds, including where the wreck has been designated a war grave. Those in deep waters may be less of a problem, but many act as magnets for unscrupulous salvage merchants who, through lack of care, can release material in addition to the grave violation. Inshore wrecks are more vulnerable and pose greater risk especially with ageing structures.

A UK example is the wreck of the SS *Richard Montgomery*, from WW II, mentioned in Chapter 6. Until recently, the safest course was to leave the ship in place, as tools for safely managing material in muddy, tidal waters did not exist.

However, movement through time now suggests that action necessary and the tools and understanding will need to be developed once the options have been assessed.

One factor that must be considered is that of climate change, with possible changes in temperature and constituents of seas or lakes. When coupled with changing weather patterns, this may affect corrosion, release, and transport, so that material may be washed up on shore unexpectedly. This will, therefore, increase the likelihood of accidental release and change the balance of risk. This will need to be monitored and managed to prevent unexpected problems developing.

11.4.1 Land Management and History

This has been extensively discussed earlier, in Chapter 3, and will continue to be a major issue. For ranges, the tools to monitor exist, but these must be used and developed to meet new requirements.

Equally, the techniques developed elsewhere can be used to assess and manage ranges where there has been no management plan till recently – assessment can be done successfully and thoroughly and thereby lead to a plan for future management.

Here too, the impact of climate change must be considered. Some immobilized material may well be released as a result of changes in temperature or rainfall, and so pose additional risks to populations [4–11].

The methods outlined were used in Bulgaria to assess a test range in the centre of the country [12]. The results indicated the level of contamination and its locations so that range management could be improved, and priorities identified. This was a highly effective programme utilizing the techniques identified in the North Atlantic Treaty Organization (NATO) studies to assess and develop a management plan for this range. It was a successful work example and indicates that the tools and techniques described earlier can be used effectively.

As time passes and new uses are sought, problems can arise with management of land, as in this example in the Scottish Borders.

Example of historical problems

- Stobs camp in Scotland
    - Large WW1 training camp and POW camp.
    - Used in WW2
    - Firing range
        - No records
        - Some finds
        - Munition possibly dumped underwater
    - Plans to develop it as a historical site
    - Assessment under way

Unplanned disposal is not likely to diminish and cleaning up is certain to remain a live issue. The year 2014 also reminded us that material from the 1914–1918 war still requires handling!

Officials in France have confirmed the discovery of an ordnance disposal site in the northern French region of Meuse, where massive numbers of shells left over after World War I were dismantled, contaminating the soil [13]. In July 2016, local authorities banned farmers from selling produce grown in the area, which had been used as farmland for several decades. The areas affected by the ban are close to Verdun, the North-Eastern French city that gave its name to the famous World War I battle in which more than 300 000 German and French soldiers died. Results of initial soil testing carried out by France's Geological and Mining Research Bureau show traces of metal and chemical compounds in the soil including arsenic, lead, and zinc. Traces of explosives and industrial chemicals used in the disposal of the shells were also found.

It is not merely recent battlefield contamination that is a problem. Recent archaeological investigation of the original Old College of Edinburgh University, replaced in the eighteenth/nineteenth century, found the chemistry laboratory used for research by Joseph Black, professor in the late eighteenth century, and heavy contamination from cobalt, mercury and arsenic from his experiments were found [14]. Residues can endure.

11.5 Security Issues and Their Impact on Requirement

As mentioned, there is a requirement to manage surplus munitions so that they are not available for misuse. This must include both unused/surplus munitions and the residues from misfirings or even from historical testing or conflict. Efficient disposal must be planned and exercised with due care. NATO Support and Procurement Agency (NSPA) is actively involved in this and there are studies under way aimed at minimizing the security threat through removal and disposal of materials.

Historical materials can pose different problems as the quantities and nature can be unknown, and as a result so can the risks – documentation is often lacking; and unusual and unfamiliar materials are used, which makes management harder. Improvised materials can pose similar problems

11.6 Future Options and Needs in a Changing Political Landscape

Environmental impact is part of the whole life of a munition and its ingredients. Experience elsewhere has shown that the whole life needs to be examined to understand and optimize the behaviour and thereby reduce the environmental impact. This formed the basis of a further study. Parts of the report are available and have been published in a series of papers [1].

At the outset of this study, several key issues that appeared to need examination were identified:

Ingredients
Manufacturing
Use
Whole life-cycle management
Disposal
Impact on environment

As noted earlier, it became clear that the concept of greener munitions is far from simple. Not only are the individual aspects more complex but their interactions are important and equally complex.

Several directions are being actively pursued at present. Some will naturally prove to be more effective than others.

Research will continue on the study of new or improved materials, but, as has been shown, there are risks associated with that, so tools are required for their assessment including prediction of toxicology as discussed in earlier chapters.

Amongst those being examined are novel materials such as co-crystals, coated materials, high energy materials based on nitrogen, and more carefully designed materials. Designer, well-characterized materials could have a significant impact on planning for the future, but the cost–benefit and overall environmental impact will need to be justified using tools such as toxicity and life-cycle estimation. These should be capable of answering planning and usage questions, although they will require refinement and improvement if they are to evolve to meet future needs.

It may be that the tools can also be used to help estimate the risks associated with existing materials, either in store or *in situ*, although this will be more challenging.

Methods for the handling of contamination will continue to develop, including the use of biological systems for enhanced natural attenuation.

New methods for more efficient manufacture are being investigated. These are aimed at cost reduction through more efficient approaches which will also lessen the environmental impact. They can also offer routes to new forms of materials. One of these, Resonant Acoustic Mixing appears to offer some major advantages for mixing, control of effluents, and the ability to make new ingredients such as co-crystals effectively [15]. A schematic of the mixing process is shown. It involves resonance mixing of ingredients in a controlled and programmable way within a closed container, and illustrates that novel approaches are being taken to provide new options. The level of publication is increasing, but there are several areas where research is needed to understand and validate the process. These include temperature management and scalability.

It must be observed that this investment is not purely for environmental reasons but that it forms part of the perceived possible advantages of the technique. Any method that offers a range of advantages is more likely to be exploited and this is likely to be true for many of the options. It will be rare for environmental requirements to monopolize the approach.

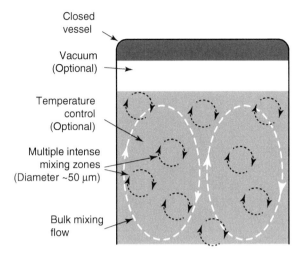

Schematic illustration of resonant acoustic mixing technology.

Other methods offering cleaner and more effective processing have been studied, such as reactive injection moulding [16] and twin screw extrusion – both illustrated. While these have been available for some time, their use may now be appropriate to reduce the environmental impact and benefit the life-cycle costs.

In summary, one listing of the Materials approach could be as follows:

Conception:
What are the predicted chemical properties?
What are the toxicity estimates?
Using computer modelling/computational chemistry to predict chemical properties and toxicity.

Synthesis:
What are the chemical/physical properties?
Screen for toxicity.
Problems with biproducts, effluent, and scaling.

Production:
Is there a potential for workplace exposure?
Will there be effluent/waste that requires disposal?

Use/demilitarization:
Are servicemen, civilians, residents potentially exposed?
Is there a potential for environmental releases (e.g. through partial detonations)?
What are the main degradation combustion and detonation products?
What is the environmental load for blow-in-place/open burning/open detonation/demilitarization/recycling?
What is the lifetime of the material?
Do techniques exist for effective disposal of recycling?

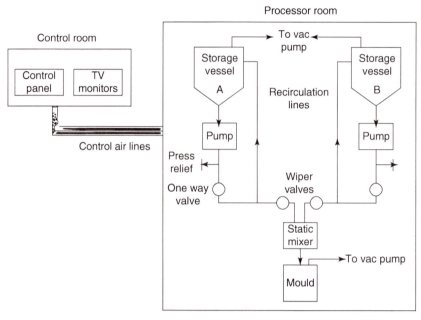

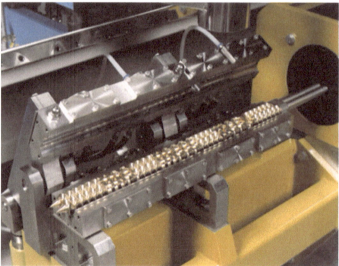

(a) Reactive injection moulding schematic. (b) Twin screw extruder.

11.7 Conclusions

This book is intended to provide an introduction to the technical area and to provide sufficient information to help manage environmental issues associated with munitions systems.

In summary, to manage the potential environmental impact of energetic systems, we need a range of approaches. Firstly, while it is not merely a matter of using new materials, they do offer sound options. However, they need to be understood well enough to deliver all the requirements placed upon them. This requires an understanding of likely toxicology and environmental and human impact as well as performance, ageing, and vulnerability. Since value for money also needs consideration, it may be that better specified and understood versions of existing materials will be more rapidly and effectively employed.

New processes can reduce manufacturing impact. Many processes were designed when there was less understanding of the effects and new approaches can be more efficient with reduced cost.

New range management methods avoid damage and remove old damage. This is not limited to test ranges but also for manufacturing plants and storage facilities.

Overall, therefore, systems design for life minimizes overall impact!

These constraints and requirements should be considered a major driver for research and a scientific and engineering challenge. They require the following:

New methods for analysis;
New or re-engineered and well-characterized materials for use;
New methods for disposal.

They also require both understanding and a rigorous approach which will minimize the risks generally.

References

1 Cumming, A.S. (2017). Energetic materials and the environment. *Propellants Explos. Pyrotech.* 42: 1–17.
2 Alliod, C., Chemelle, J.-A., Jacob, G., and Terreux, R. (2017). Ames testing prediction on high energy molecules by on-the-fly QSAR (OTF-QSAR). *Propellants Explos. Pyrotech.* 42: 24–35.
3 NATO RTO (2010). Environmental Impact of Munition and Propellant Disposal. *RTO-TR-AVT-115*, open publication.
4 Walsh, M.R. et al. (2012). Munitions propellants residue deposition rates on military training ranges. *Propellants Explos. Pyrotech.* 37: 393–406.
5 Thiboutot, S. Characterisation of Residues from the Detonation of Insensitive Munitions. *SERDP ER-2219*. https://www.serdp-estcp.org/Program-Areas/Environmental-Restoration/Contaminants-on-Ranges/Characterizing-Fate-and-Transport/ER-2219/(language)/eng-US.
6 Walsh, M.R., Walsh, M.E., Ramsey, C.A. et al. (2014). Energetic residues from the detonation of IMX-104 insensitive munitions. *Propellants Explos. Pyrotech.* 39: 243–250.
7 Walsh, M.R., Walsh, M.E., Poulin, I. et al. (2011). Energetic residues from the detonation of common US ordinance. *Int. J. Energetic Mater. Chem. Propul.* 10 (2): 169–186.

8 Ruppert, W.H. et al. (2010). A history of environmentally sustainable energetics. In: *Insensitive Munitions and Energetic Technology Symposium*. USA: NDIA.
9 Brinck, T. (2014). *Green Energetic Materials*. Wiley.
10 Sabatini, J.J. and Shaw, A.P. (2014). Advances toward the development of "Green" pyrotechnics. In: *Green Energetic Materials*. NATO STO AVT 243 Proceedings.
11 Sunahara, G. (ed.) (2009). *Ecotoxicology of Explosives*. Wiley.
12 Hristov, H.P. and Hristov, H.I. (2017). Environmental impact of energetics on test ranges. *Propellants Explos. Pyrotech.* 42: 84–89.
13 Auborg, L. (2015). https://news.vice.com/article/how-shells-from-world-war-i-may-be-contaminating-french-food-today.
14 Grier, J. and Bownes, M. (2016). *Digging Up Our Past – The Archaeological Excavations in Old College Quad*. University of Edinburgh.
15 Hope, K.S., Lloyd, H.J., Ward, D. et al. (2015). Resonant acoustic mixing: its application to energetic materials. In: *Proceedings of New Trends in Research in Energetic Materials*, 134. Czech Republic.
16 Flower, P. and Marshall, E.J. (1993). Comparison of plastic bonded explosives produced by reaction injection moulding and conventional production. In: *Proceedings of 24th ICT*, 53. Pfinztal, Germany: Fraunhofer ICT.

Index

a

acaroid resin   114
acoustic sensors   145–146
afterburning   33, 34
aftermath–residues   2
air entrainment   33, 34
2-amino-4,6-dinitrotoluene (2-A-DNT)   51
4-amino-2,6-dinitrotoluene (4-A-DNT)   51
ammonium dinitramide (ADN)   5, 80
ammonium perchlorate (AP) oxidizer   51, 70, 199
ammunition demilitarization   13–45, 176
anti-tank 'Barmines'   36
antitank ranges   60, 68
aqueous solubility (S_w)   58
artillery guns   66, 79
autonomous underwater vehicles (AUVs)   146

b

bathymetry   146
$B_4C/KNO_3/KCl/Ca$ stearate/PVAc   125
beeswax   114
Best Available Techniques Not Entailing Excessive Costs (BATNEEC) principle   43–44
binders   26, 51, 96, 103, 104, 114, 117, 120, 124
bioaccumulation factor (BCF)   218
bioturbation   143, 154
5,5'-bis(1-hydroxytetrazole)   122
blow-in-place (BIP) detonations   59, 78
bulk propellants   32, 33

c

C4   92–97
calibre projectiles   64
cannon ammunition   22, 23
carbon reduction   171
characterization factors (CF)   184, 186, 219, 220, 224
Chemical Munitions Search and Assessment (CHEMSEA)   150, 157
Chemical Safety Assessment (CSA)   215
Chemical Safety Report (CSR)   216
chemical warfare agents (CWA)   139, 156, 160
Chemring Nobel Company   96
chlorobenzenes   108
chlorophenols   108
circular economy   171
civil authorities   1
civilian pyrotechnic systems   107
clean-up strategy   105
CO_2 estimates   38
Cold Regions Research and Engineering Laboratory (CRREL)   47
commercial demilitarization contract   15
commercially viable materials   19
Composition B (Comp B)   51, 69, 77, 82
contaminated water   58, 227
contamination management   7
corrosion   63, 64, 140–145, 151, 155–158, 233, 246
cost-efficiency analysis (CEA)   83
costs and associated logistics   15
Council Directive 67/548/EEC   214
CX-85   83–85, 89

Energetic Materials and Munitions: Life Cycle Management, Environmental Impact and Demilitarization,
First Edition. Edited by Adam S. Cumming and Mark S. Johnson.
© 2019 Wiley-VCH Verlag GmbH & Co. KGaA. Published 2019 by Wiley-VCH Verlag GmbH & Co. KGaA.

d

data quality objective (DQO) process 53
Decision aid for marine munitions (DAIMON) 141
decision units (DUs) 53
deep sea dumping 13, 17
Defence Research and Development Canada (DRDC) 47
demilitarization 235
  basic stages of 17–20
  examples of cost and CO_2 36
  facilities 20
  factors influencing 15–17
  maturity and use of 21–26
  munition manager's perspective 44–45
  NATO AOP 4518 40–44
  open burning (OB) 29–33
  open detonation (OD) 29–31, 33–36
  scale of issue 14–15
  technical and environmental issues 27–29
  techniques and processes 20–21
Department of National Defence Director General Environment (DGE) 47
design for demilitarization (DFD) 235
  munition manager's perspective 44–45
  NATO AOP 4518 40–44
design for disposal 173
detonation 2, 33–34
  residues 90–91, 233
detonators 32, 33, 94, 132, 243
3,3'-diamino-4,4'-dinitramino-5,5'-bi-1,2,4-triazolate 121
1,1-diamino-2,2-dinitroethylene (FOX-7) 80
diazodinitrophenol (DDNP) 181
dibutyl phthalate 105, 213, 220–223
diisononylphthalate 220
2,4-dinitotoluene 231
2,6-dinitroanisole 231
dinitroanisole (DNAN) 69, 80
4,5-dinitro-1,3-imidazole 121
2,4-dinitrotoluene (2,4-DNT) 51
2,6-dinitrotoluene (2,6-DNT) 51
dinitrotriazalone (NTO) 80
dioctylsebacate 220
dioctyl terephthalate 220–224
dioxin/dibenzofuran cocktail of pollutants 108
dioxins 108
Directive 76/769/EEC 214
Directive 79/831/EEC 214
Director Land Environment (DLE) 47
disposal and waste burning 126–127
disposed military munitions (DMM) 139
DM12 detonation and deposition rate measurement 94
2,4-DNT 51, 52, 58, 59, 64–66, 78, 80, 81, 88, 89, 152
Doppler velocity log (DVL) 148
DRDC Valcartier 48, 61, 90

e

ecodesign 171, 173, 174, 194
Ecodesign Directive 173
ecoinvent database 179, 187
ecoinvent website 179
ecolabel 171
ecotoxicity 186, 187, 190, 192, 204, 220, 224, 237
ecotoxicology 8
effect factor 186
embedded electronics 41
EM sensors 146
end-of-life (EOL) disposal 41
end-of-mission (EOM) disposal 41
end-of-operational-life (EOOL) 41
Engineer Research and Development Center (ERDC) 47
environmental and safety legislation 213
environmental assessment 2, 16, 28, 38, 39, 64, 180, 200
environmental hazard assessment 1
environmental impact of munitions 2, 247
environmental life-cycle impacts 188, 191
Environmentally Sustainable Manufacturing for Energetic Formulations 243
environmentally sustainable RTAs 58
environmental management 1, 20, 68, 176, 194, 225, 243
Environmental Management System (EMS) ISO 14001 27
Environmental Occupational Health and Safety Assessment in Canada 63

Environmental Protection for Heavy
  Weapons Ranges (EPHW) 47
environmental releases 200, 204, 227,
  231, 237, 249
environmental safety and occupational
  health (ESOH)
  acquisition 233–234
  conception 228–232
  cost and time 208–210
  current and evolving regulatory
    interests 207
  data requirements 201–207
  decision-making 238
  demilitarization 235
  disposal 237
  ecotoxicity 237
  engineering and
    manufacturing 234–235
  evolving science and new tools 203
  fate and transport 235–237
  field monitoring 237
  industrial hygiene 237–238
  integration of flow charts 204
  life cycle environmental
    assessment 200
  material synthesis 231
  M116, 117, 118 simulators 207–208
  munition compounds and aetiology
    of 199–200
  M-18 violet smoke 208
  phased approach 201–203
  Registration, Evaluation, Authorisation,
    and Restriction of Chemicals
    (REACH) 238
  reproduction/developmental
    effects 204–206
  research and acquisition 200–201
  research vs. testing 203–204
  testing/demonstration 232–233
  toxicology data requirement 228
  Toxic Substances Control Act 238
Environmental Security Technology
  Certification Program
  (ESTCP) 104, 159
EPA 8330b method 56, 68
European Chemical Agency
  (ECHA) 215, 238
European Conference on Defence and the
  Environment (ECDE) 48

explosive footprints 61–64
explosive hazard 18, 20, 27
explosive ordnance disposal (EOD) 15,
  53, 92, 235
explosive waste incinerator (EWI) 22,
  23, 28
exposure factor 186, 203

f
fate factor 186
Finnish Defence Administration 48
firing positions (FP) 51, 54, 59, 64–66,
  68, 75, 78, 79, 91
fluffy layer of suspended matter
  (FLSM) 144
four-power programme 5
Free From Explosive Hazard 27
freshwater ecotoxicity 185
fuze 18, 82, 83, 92, 139
fuzeheads 117

g
GIM 83–85, 89–91
  explosive residues 91
Global Navigation Satellite System
  (GNSS) 148
GLObalnaya NAvigatsionnaya
  Sputnikovaya Sistema
  (GLONASS) 148
global warming 170, 183, 186, 189
greener munitions
  definition 79, 242–243
  development approach 79–82
  green plastic explosive 92
  hexahydro-1,3,5-trinitro-1,3,5-triazine
    (RDX) 77
  IM 60-mm mortar based on
    PAX-21 77
  M115A2 ground burst projectile
    simulator 77
  M116A1 hand grenade simulator
    77
  M274 2.75" rocket simulator 77
  munitions constituents of
    concern 77–78
  RIGHTTRAC 82–92
  source of munitions
    constituents 78–79
  unexploded ordnances (UXO) 75

green plastic explosive
  HMX option   95–96
  PETN option   93–95
green procurement   171, 173
green pyrotechnic systems   104
groundwater plume contamination   49
gun propellant   51, 64, 79, 82, 83, 86, 88, 213, 220, 243
gun testing   89–90

h
Hawaii Undersea Military Munitions Assessment (HUMMA)   142, 143, 157
hazardous wastes   14, 18, 20, 28, 33, 56, 127, 199, 234
  industrial demilitarisation   19
Health and Safety Executive (HSE) Explosives Industry Group   35
heavy metals   18, 49, 77, 104, 105, 108, 109, 115, 118–122, 155, 169, 170, 184
hexachloroethane (HCE) smoke systems   116
1,3,5-hexahydro-1,3,5-trinitrate (RDX)   199
hexahydro-1,3,5-trinitro-1,3,5-triazine (RDX)   77, 152
2,4,6,8,10,12-hexanitro-2,4,6,8,10,12-hexaazaisowurtzitane (CL-20)   80
hexavalent chromium compounds   115
HK416   76
human health and ecosystems   169, 184, 185, 187, 213–215, 218–220, 225
human health (CTUh) and ecosystems (CTUe)   186
human-occupied vehicles (HOVs)   142
human toxicity   186, 187, 189, 192, 193, 220, 224
hydrolysis   58, 141, 144, 153, 229, 232

i
IM 60-mm mortar based on PAX-21   77
individual disposal/demilitarisation action   13
industrial demilitarization   13, 17, 19, 21, 22, 27, 30, 31, 36, 39, 44
Industrial Emissions Directive   28
industrial hygiene   208, 227, 234, 237–238
inertial navigation systems (INS)   148
input–output analysis   171
insensitive high explosive (IHE)   41, 69, 70
insensitive munitions (IM)   6, 70, 77, 82, 169
Intergovernmental Panel on Climate Change (IPCC)   185
in vitro–in vivo extrapolation (IVIVE)   207
ISO 9001   27
isocyanates   6, 117

j
just-in time manufacturing   111

l
L320   86–90
land management   246–247
large munitions   44
large-scale explosive waste incinerator   23
laser line scanners (LLS)   146
lead azide   105, 106
lead-free bullets/primers   76
lead styphnate (TNR-Pb)   106, 181
legislative impact   2–4
life cycle analysis   7, 91–92, 199
life cycle assessment (LCA)
  methodology   171–172
  environmental and toxicological impacts   170
  four interrelated phases   172
  functional unit   175–176
  goal and scope phase   174–178
  ISO standards 14040 and 14044   172
  life-cycle impact assessment (LCIA)   182–194
  life-cycle inventory   178–182
  life-cycle thinking   170–171
  limitations of   194
  purpose of   173–174
  simplified representation   171
life-cycle costing   171
life cycle environmental assessment   200
life-cycle impact assessment (LCIA)
  case study   188–194
  characterization   183

classification 183
methods 185–187
normalization and weighting 183
software 187–188
life-cycle inventory (LCI) 172, 178–182
local exhaust ventilation (LEV) systems 117
long baseline (LBL) 148

m

magnesium teflon viton (MTV) 116–117
M115A2 ground burst projectile simulator 77
M116A1 hand grenade simulator 77
marine Mk144 illumination signal 122
material flow analysis 171
maximum acceptable concentrations (MACh) 58
metal-based colourant 113
methyl centralite (MC) 88, 89
military live-fire training ranges
   analytical tool and adsorption method for MC in aqueous samples 67–68
   DRDC Valcartier 48
   emerging constituents 69–70
   explosive footprints 61–64
   firing positions 64–66
   groundwater plume contamination 49
   mitigation measures 67–69
   munition related contaminants 51
   octahydro-1,3,5,7-tetranitro-1,3,5,7-tetrazocine (HMX) 48
   surface soil characterization 52
   thermal treatment of shoulder rocket propellant contaminated surface and sub-surface soils 68–69
   unexploded ordnance (UXO) 50
105-mm army artillery munition 82
105-mm Howitzer gun 89, 90
155-mm artillery munition (M777) 87
mobile demilitarisation plants 40
modified single-base (MSB) 86–89
Monitoring of dumped munitions (MODUM) 151, 154
M274 2.75" rocket simulator 77

M116, 117, 118 simulators 207–208
multicriteria analysis 171
Multiple Launch Rocket Systems (MLRS) 38
munition constituents (MC) 52, 225
munition manager's perspective 44–45
munition-related contaminants 51–52
munitions, handling of 151–152
munitions and explosives of concern (MEC) 139
   acoustic sensors 145–146
   chemical degradation 152–153
   corrosion 142–143
   detection 150–151
   ecotoxicological aspects 154–156
   EM sensors 146
   environmental aspects 141–142
   fate and transport of constituents 144–145
   geopolitical aspects 156–158
   global collaboration 159–160
   global EU and NATO efforts 160–161
   handling 151–152
   human capacities 161–162
   in situ methods 141
   location 150
   long-term and long-distance transport 153–154
   monitoring 151
   navigation and positioning systems 148
   optical sensors 146
   platforms 146–147
   remediation 141–150
   research infrastructures 161
   scientific advances 161
   sea disposal process 145
   technology innovation 161
munitions constituents (MCs) 47
   of concern 77–78
   RTA 58–61
   source of 78–79
munitions, disassembly of 13, 27
Munitions Safety Information and Analysis Centre (MSIAC) 45
M-18 violet smoke 208

n

NAMMO Buck 29, 30
NAMMO NAD 29
national legislation and public acceptance 15
NATO AOP 4518 40–44
NATO Applied Vehicle Technology (AVT) 48
NATO AVT 115 4
NATO AVT 177 4
NATO AVT 179 4
NATO AVT 197 4
NATO AVT 269 4
NATO AVT-249 task group 48
NATO Collaboration Support Office tasked AVT-197 48
NATO Cooperative Demonstration of Technology (CDT) 48
NATO Industrial Advisory Group (NIAG) 21
NATO RTO AVT-177 symposium 63
NATO Support Agency 8
NATO Support and Procurement Agency (NSPA) 16, 247
navigation and positioning systems 148
neurotoxin 6
N-guanylurea-dinitramide (FOX-12) 80
nitramines 51, 80, 95, 231
nitrate esters 7, 51, 231
nitroaromatics 81, 153, 231
nitroaromatic-trinitrotoluene (TNT) 2, 4, 6, 80, 152
nitroglycerin (NG) 51, 59, 69, 97, 231
nitroguanidine (NQ) 51, 59, 86
nitrotriazalone (NTO) 69, 80
North American RTAs 49, 51, 82
Norwegian Armed Forces 76
Norwegian servicemen 6

o

obscurant smokes 124
octahydro-1,3,5,7-tetranitro-1,3,5,7-tetrazocine (HMX) 48, 59, 80, 95, 152
octanol/water partition coefficients (K_{ow}) 58, 209
octol (TNT/HMX) 51
OHSAS 18001 27
old red lead/silicon primer 115
One health approach 199–210
on-site disruption and disposal 8
open burning (OB) 4, 22, 25, 29–33, 66, 78, 128, 170, 235, 249
 vs. EWI incineration of SAA 39
open detonation (OD) 4, 13, 22, 25, 29–31, 33–36, 52, 170, 176
optical sensors 145, 146
ordnance disposal site 247
Organization for Security and Cooperation in Europe (OSCE) 23
oxydiethane-2,1-diyl dibenzoate 220, 224
ozone depletion 183, 184

p

pentaerythritol trinitrate (PETN) 80
  option 93
perchlorates 7, 49, 51, 59, 77, 78, 97, 104, 108, 109, 112, 120, 199
  chlorate 116, 122–124
periodates 120, 123
persistent, bioaccumulative, or toxic (PBT) 216
phosphorus (V) nitride (P3N5) 118
physical destruction techniques 20
plastic-bonded explosives (PBX) 41
plume dispersion 33, 34
plume formation 33, 34
polychlorinated biphenyls 123
post-detonation residues 62, 77, 91, 93
Prevention of Marine Pollution by Dumping of Wastes 139
Programmatic Environmental Safety and Health Evaluation (PESHE) 200
propellant production 192, 220
propellant residues 54, 59, 60, 65, 66, 68, 79, 236
propellant's environmental hazard assessment 1
propelling charge system 66
pyrotechnics 5, 227
  civilian pyrotechnic systems 107
  'clean-up' strategy 105
  compositions 103, 104
  disposal and waste burning 126–127

environmental effect   103, 104
environmental legislation   127, 129
green pyrotechnic systems   104
heavy metal   115, 118–122
integration   129–133
list of ingredients   104
magnesium teflon viton (MTV) countermeasures   116–117, 127
obscurant smokes   124
packaging waste   118
perchlorate and chlorate   116, 122–124
perchlorate levels   104
production   109–110
pyrotechnic devices   112
qualification   107
raw materials acquisition and quality control   112–114
Registration, Evaluation, Authorisation and Restriction of Chemicals (REACH)   105–106
resins, binders and solvents   117
and simulators   32
site location   110–112
smokes   116
specific materials production   114–115
storage   117–118
suitably qualified and experienced person (SQEP) issues   128–129
usage and disposal   118
volatilization smokes   116, 124–125

q

quality assurance/quality control (QA/QC)   57, 231
quantitative structural activity relationship (QSAR) models   230
quantitative structural property relationship (QSPR)   230

r

range and training areas (RTA) characterization   47
R440 dim illuminant   122
RDX   6, 49, 51, 52, 58–60, 63, 68, 77, 79, 80, 82, 83, 85, 92–97, 149, 152–156, 182, 199, 213, 231
RDX/TNT   36, 37

reactive injection moulding   249, 250
Registration, Evaluation, Authorisation and Restriction of Chemicals (REACH)   2, 105, 238
  European defence capabilities   213
  overview of   214–215
  regulation 793/93   214
  regulatory thresholds   218
remotely operated vehicles (ROVs)   142
removal from storage   17–18
residual material disposition   20
resins   114, 117
resonant acoustic mixing technology   249
reuse, recovery and recycling (R3) methods   18, 26, 41
Revolutionary Insensitive, Green and Healthier Training Technology with Reduced Adverse Contamination (RIGHTTRAC)
  cost-efficiency analysis (CEA)   83, 91
  fate, transport and toxicity   88–89
  field demonstration   89
  IM properties   83–84, 87–88
  main explosive charge   83
  105-mm army artillery munition   82
  performance   83
  propellant charge   86
rising gas   143
rocket motors   18, 26, 29, 32, 33, 45, 243

s

Safe Drinking Water Act   97
safety data sheet (SDS)   106
sea disposal   139, 140, 145, 146
sea disposal process   145
sea-disposed munitions   140, 143, 161
sea-dumped munitions   4, 160, 245
sea dumping   13, 159, 176, 241
SENTINEL project   107
settling sediments   143
short baseline (SBL)   148
shoulder rocket propellant contaminated surface and sub-surface soils   68–69
SimaPro   187, 188
single-base propellant   51, 86, 87
size reduction/removal   18
small arms ammunition (SAA)   15, 28, 29, 32, 33, 39, 40

Index

small arms ammunition destruction oven (SAADO) 33
small arms propellant residues 65
small-scale incinerator and products 22
smokeless propellant 5
snow sample filtration 62
soil sorption constants (K_d) 58
solvents 32, 56, 105, 116, 117, 125, 243
space agencies 1
SS Richard Montgomery 158, 245
steel/lead projectile 188–193
Strategic Environmental Research and Development Program (SERDP) 47, 77, 104, 159, 210, 239, 243
styphnate 105, 106, 152
substances of very high concern (SVHC) 105, 213, 214, 216
substantial residues 170
suitably qualified and experienced person (SQEP) issues 128–129
surface soil characterization
 cleaning 57–58
 data quality and sampling objectives 53–55
 risk to the receptors through the transport of munitions constituents 58–61
 safety aspects 53
 soil samples 56–57
surface soil sampling pattern 55
surface to near-surface UXOs 64
surface vessels 146–148
system boundaries 172, 175–178

t

Technology Demonstration Program (TDP) 77
terrestrial toxicity 80, 85, 88
thermal treatment 32, 68–69, 170
toxicological life-cycle impacts 189
Toxic Substances Control Act 238
trichloroethane (TCE) 105
1,3,5-trinitrobenzene (TNB) 58, 154–156, 231

2,4,6-trinitrotoluene (TNT) 6, 51, 59, 77, 80, 152, 156, 176, 200, 231
twin screw extrusion 249

u

ultrashort baseline (USBL) 148
unexploded ordnance (UXO) 35, 50, 75, 92, 139, 170
United Nations Environment Programme/ Society of Environmental Toxicology and Chemistry (UNEP/ SETAC) 185
unplanned disposal 8, 247
unmanned surface vehicles (USVs) 147
unmanned vehicles 146
unused/surplus munitions 247
USA peer-reviewed funding program 47
U.S. Army Corps of Engineers 6
U.S. Army Environmental Quality Technology Program 208
U.S. Army's Green Ammunition Program 76
US Environmental Protection Agency 2, 49, 78, 104, 201, 238
U.S. EPA Contaminant Candidate List 3 (USEPA web site) 97
U.S. EPA Interim Lifetime Drinking Water Health Advisory 78
USEtox method 185–189, 193, 194, 218–220, 224
U.S. Massachusetts Military Reservation (MMR) 49, 76
US Strategic Environmental Research and Development Program (SERDP) 47, 77, 104, 159, 243

v

volatilization smokes 116, 124–125

w

Wallop Defence Systems site 117
waste disposal 19, 20, 110, 126
waste management 110–112, 116, 126, 176
watercraft 147
water solubility of energetic materials 81
'whistle, bang, flash' simulators 207–208